AF821984

Astronomy and Chemistry for the Logic Stage

Student Guide

THIS BOOK IS A CUSTOM SCHOOL EDITION

Please be aware that it will not match any of the current editions of the programs on our website. If you need additional copies, please contact us at support@elementalscience.com with your school name.

The images and all other content in this book are copyrighted material owned by Elemental Science, Inc. Please do not reproduce this content on e-mail lists or websites. Duplicating this book or printing the eBook so that the book can then be reused or resold is a violation of copyright.

Astronomy and Chemistry for the Logic Stage Student Guide

Custom School Edition, Valor Preparatory Academy
Email: support@elementalscience.com

ISBN: 978-1-935614-97-5

For more copies write to:
Elemental Science
PO Box 79
Niceville, FL 32588
support@elementalscience.com

Copyright Policy

Astronomy and Chemistry for the Logic Stage
Student Guide Table of Contents

Astronomy and Chemistry for the Logic Stage
Letter to the Student

Dear Student,

Welcome to your journey through astronomy and chemistry. Astronomy is the study of space, while chemistry is the study the various forms of matter, their composition, structure, and properties, as well as how they react with each other. This year, you will examine our solar system and what can be found in space, along with atoms and molecules along with how they are related to the various forms of energy. You will also look at the building blocks of chemistry, matter, solutions, mixtures, acids, and bases along your voyage. This guide is written for you, so enjoy your journey!

What this guide contains

First, this guide includes your date sheets and unit sheets. The unit sheets have your vocabulary words, weekly student assignment sheets, sketches, experiment sheets, and space for each of your writing assignments. After your unit sheets, you will find the Appendix. In it, you will find a list of all your memory work for the year, a glossary, and a place to record any additional activities you have done which pertain to chemistry.

Student Assignment Sheets Explained

The Student Assignment Sheets contain your weekly assignments for each week. Each of the student assignment sheets contains the following:

- ✓ **Experiment** — Each week will revolve around a weekly topic. You will be assigned an experiment to complete that poses a question about what you are studying. Each student assignment sheet contains the list of materials you will need and the instructions to complete the experiment. This guide also includes experiment sheets for you to fill out each week. In each of these experiments, you will use the scientific method.

  A Word about the Scientific Method —The scientific method is a method for asking and answering scientific questions. This is done through observation and experimentation. The following steps are key to the scientific method:

  1. **Ask a Question** — The scientific method begins with asking a question about something you observe. Your questions must be about something you can measure. Good questions begin with how, what, when, who, which, why or where.
  2. **Do Some Research** — You need to read about the topic from your question so that you can have background knowledge of the topic. This will keep you from repeating mistakes of the past.
  3. **Formulate a Hypothesis** — A hypothesis is an educated guess about the answer to your question. Your hypothesis must be easy to measure and answer the original question you asked.
  4. **Test with Experimentation** — Your experiment tests whether your hypothesis is true or false. It is important for your test to be fair. This means that you may need

to run multiple tests. If you do, be sure to only change one factor at a time so that you can determine which factor is causing the difference.

5. **Record and Analyze Observations or Results** — Once your experiment is complete, you will collect and measure all your data to see if your hypothesis is true or false. Scientists often find that their hypothesis was false. If this is the case, they will formulate a new hypothesis and begin the process again until they are able to answer their question.
6. **Draw a Conclusion** — Once you have analyzed your results, you can make a statement about them. This statement communicates your results to others.

Each of your experiment sheets will begin with a question and an introduction. The introduction will give you some background knowledge for the experiment. The experiment sheet also contains sections for the materials, a hypothesis, a procedure, an observation and a conclusion. In the materials section you need to fill out what you used to complete the experiment. In the hypothesis section you need to predict what the answer to the question posed in the lab is. In the procedure section you need to write step by step what you did during your experiment (so that someone else could read your report and replicate your experiment). In the observation section you need to write what you saw. Finally, in the conclusion section you need to write whether or not your hypothesis was correct and any additional information you have learned from the experiment. If your hypothesis was not correct, discuss why with your teacher and then include why your experiment did not work on your experiment sheet.

Safety Advisory: Do not perform any of the experiments marked "☢ **CAUTION**" on your own. Be sure you have adult supervision.

☐ **Vocabulary & Memory Work** —Throughout the year, you will be assigned vocabulary and memory work for each unit. Each week, you will need to look up the word in your glossary and fill out the definitions on the Unit Vocabulary sheet found at the beginning of each unit in this guide. You may also want to make flash cards to help you work on memorizing these words. Each week, you will also have a memory work selection to work on. Simply repeat this selection until you have it memorized, and then say the selection to your teacher. There is a complete listing of the memory work selections in the Appendix on pp. 323-326 of this guide.

Sketch — Each week, you will be assigned a sketch to complete. Color the sketch and label it with the information given on the Student Assignment sheet. Be sure to give your sketch a title.

Writing — Each week, you will be writing an outline and/or a narrative summary. The student assignment page will give you a reading assignment for the topic from your spine text. After you have finished the assignment, discuss what you have read with your teacher. Your teacher will let you know whether to write an outline or a narrative summary from your spine text reading. Your teacher may also assign additional research reading out of the additional books.

Once you finish the additional reading, prepare a narrative summary about what you have learned from your reading. Your outlines should be one-level main topic style outlines, and your narrative summaries should be one to three paragraphs in length, unless otherwise assigned by your teacher.

- **Dates** —Each week, dates of important discoveries within the topic and dates from the readings are given on the student assignment sheet. You will enter these dates onto one of four date sheets. The date sheets are divided into the four time periods laid out in *The Well-Trained Mind* by Susan Wise Bauer and Jessie Wise (Ancients, Medieval-Early Renaissance, Late Renaissance-Early Modern, Modern). These sheets are found in the ongoing projects section of this guide. You can choose to just write the dates and information on the sheet or if you want you can draw a timeline in the space provided and enter your dates on that.

How to schedule this study

This program is designed to take up to three hours per week. You, along with your teacher, can choose whether to complete the work over five days or over two days. Below are two options for scheduling to give you an idea of how you can schedule your week.

- ✓ A typical two day schedule:
  - **Day 1** — Define the vocabulary, do the experiment and complete the experiment page, record the dates;
  - **Day 2** — Read assigned pages and discuss together, prepare the science report or outline, complete the sketch.
- ✓ A typical five day schedule:
  - **Day 1** — Do the experiment and complete the experiment page;
  - **Day 2** — Record the dates and define the vocabulary;
  - **Day 3** — Read assigned pages and discuss together and complete the sketch;
  - **Day 4** — Prepare the science report or outline;
  - **Day 5** — Complete one of the Want More activities from the Teacher Guide.

Final Thoughts

As the author and publisher of this curriculum, I encourage you to contact me with any questions or problems that you might have concerning this program at support@elementalscience.com. I will be more than happy to answer them as soon as I am able. I hope that you will enjoy this program!

Sincerely,
Paige Hudson, BS Biochemistry, Author

Ancients 5000 BC-400 AD

Medieval-Early Renaissance 400AD-1600AD

Late Renaissance-Early Modern 1600 AD-1850 AD

Late Renaissance-Early Modern 1600 AD-1850 AD

Modern 1850 AD-Present

Modern 1850 AD-Present

Unit 1

Space

Unit 1: Space
Vocabulary Sheet

Define the following terms as they are assigned on your Student Assignment Sheet.

1. Universe – ____
____
____
2. Galaxy – ____
____
____
3. Cluster – ____
____
____
4. Supercluster – ____
____
____
5. Star – ____
____
____
6. Nebulae – ____
____
____
7. Black hole – ____
____
____
8. Constellation – ____
____
____

9. Planetarium – ____

____

____

10. Zodiac – ____

____

____

Student Assignment Sheet Astronomy Lesson 1
Universe

Experiment: Can I calculate the speed of light using a microwave?

Materials
- ✓ Large chocolate bar
- ✓ Large plate
- ✓ Microwave
- ✓ Ruler

CAUTION
Melted chocolate is very hot and can burn, DO NOT touch!

Procedure
1. Read the introduction to this experiment and answer the question.
2. Unwrap the chocolate bar and place it on the plate. Remove the turntable from your microwave and set the plate inside.
3. Begin by heating the chocolate bar for 1 minute; check to see if it has begun to melt. (*You are looking to see if there are two spots that have begun to melt. You DO NOT want the entire chocolate bar to melt.*) If the bar has not begun to melt, continue to heat it for 30-second intervals, checking each time to see if melting has begun.
4. Once melting has begun, carefully remove the plate from the microwave and measure the distance between the centers of the two melted points.
5. Record the distance on your experiment sheet and complete the calculations.
6. Draw conclusions and complete your experiment sheet.

Vocabulary & Memory Work
- ☐ Vocabulary: universe
- ☐ Memory Work – Work on memorizing the Types of Stars.

Sketch Assignment: Contents of the Universe
- Label the following: Earth, Our Solar System, Milky Way Galaxy, Cluster of Galaxies, The Universe

Writing Assignment
- Reading Assignment: *Kingfisher Science Encyclopedia* pg. 386-387 The Universe
- Additional Research Readings
  - Big Bang Theory: *KSE* pg. 388-389
  - The Universe: *USE* pg. 154-155

Dates to Enter
- 1929 – Edwin Hubble proves that the universe is expanding.
- 1965 – Scientists find heat waves in the universe that they believe are leftover from a vast explosion.
- 1992 – The satellite *Cosmic Background Explorer* traces background radiation and ripples in the universe that are thought to be leftover from the Big Bang.

Sketch Assignment Lesson 1

Experiment: Can I calculate the speed of light using a microwave?

Introduction

The universe is the vast expanse of space in which all things are found. It contains us, our planet, our solar system, our galaxy and at least 100 billion other galaxies. Astronomers have studied what they call the observable universe since Galileo invented the first telescope in 1609. As astronomers began to record their observations, they found the need to calculate distances, so that they could give approximate locations for the objects in the universe. Since the distances in the universe are so large, astronomers began to use light years to calculate them. A light year is the distance a wavelength of light will travel in one year, or about 5.88 trillion miles. In this experiment, you are going to try to calculate the speed of light, which scientists then multiply by the time in one year to calculate a light year.

Hypothesis

Can I calculate the speed of light using a microwave? Yes No

Materials

_______________ _______________

_______________ _______________

_______________ _______________

_______________ _______________

_______________ _______________

Procedure

____

____

____

____

____

____

____

____

Observations

Distance from the center of one melted chocolate spot to the center of the next melted chocolate spot	
in cm (centimeters)	
in m (meters)	

____

____

____

____

____

Results

Calculation #1: ____________________ x 2 = ____________________

(distance measured in m) (wavelength in m)

Calculation #2: ____________________ x 1,000,000 (Hz/MHz) = ____________________

(Mega Hertz(MHz), from the back of the microwave) (Cycles per second in Hz)

Calculation #3: ____________________ x ____________________ = ____________________

(wavelength in m) (Cycles per second in Hz) (speed of light m/s)

Conclusion

____

____

____

____

Written Assignment Lesson 1

Discussion Questions

1. How can we see other galaxies?
2. How do astronomers measure distances in the universe?
3. What does everything in the universe do?

Written Assignment Lesson 1

Student Assignment Sheet Astronomy Lesson 2
Galaxies

Experiment: Identifying Galaxies

Materials

- ✓ Galaxy cards (Your instructor will provide these cards.)

Procedure

1. Read the introduction to this experiment and answer the question.
2. Read "Four Types of Galaxies" found on pg. 327 of the Appendix. Then, fill in the Galaxy Information Chart on your experiment sheet with the information.
3. Next, look at each of the galaxy cards, determine which type of galaxy they are by using the information included, and then check your answers with your teacher.
4. Write what you have learned in the conclusion section of your experiment sheet.

Vocabulary & Memory Work

- ☐ Vocabulary: galaxy, cluster, supercluster
- ☐ Memory Work – Continue to work on memorizing the Types of Stars.

Sketch Assignment: 4 Types of Galaxies

- Label the following: irregular galaxy, elliptical galaxy, spiral galaxy and barred spiral galaxy

Writing Assignment

- Reading Assignment: *Kingfisher Science Encyclopedia* pg. 390-391 Galaxies
- Additional Research Readings
  - Galaxies: *USE* pg. 156-157
  - Our Galaxy and Beyond: *DK Astro* pg. 62-63

Dates to Enter

- 1784 – Charles Messier finds several blurry objects that he records as nebulae. These are later discovered to be galaxies.
- 1845 – Lord Rosse draws the galaxy M51, without knowing what it is.
- 1924 – Edwin Hubble presents the first evidence of other galaxies.

Sketch Assignment Lesson 2

Experiment: Identifying Galaxies

Introduction

All galaxies are massive celestial bodies made up of stars, gas and dust. They are held together by gravity and classified by their shape. There are four main types of galaxy shapes, spiral, elliptical, barred spiral, and irregular. In this experiment, you are going to read more about each type of galaxy shape and then identify several galaxies using what you have learned.

Conclusion

Galaxy Information Chart

Type of Galaxy				
Shape				
Presence of Bulges & Discs				
Presence of Gas & Dust				
Types of Stars				
Examples (from Galaxy Cards)				

Written Assignment Lesson 2

Discussion Questions

1. What is the name of the galaxy our solar system is in?
2. How are galaxies classified?
3. What are two things that astronomers believe help to determine the shape of a galaxy?
4. Describe the typical active galaxy.
5. Which cluster of galaxies does the Milky Way belong to?

Written Assignment Lesson 2

Student Assignment Sheet Astronomy Lesson 3
Stars

Experiment: Why do stars twinkle?

Materials
- ✓ Aluminum foil
- ✓ Large piece of cardboard (large enough to fit under the glass bowl)
- ✓ Glass bowl
- ✓ Flashlight
- ✓ Scissors

Procedure
1. Read the introduction to this experiment and answer the question.
2. Cut out small pieces of aluminum foil, shape them into stars and glue them onto the cardboard.
3. Fill up two thirds of the bowl with water and place it on top of the cardboard.
4. Turn off the lights in the room you are in and shine your flashlight on the top of the bowl.
5. Tap the side of the bowl and record what happens.
6. Discuss with your teacher, draw conclusions, and finish your experiment sheet.

Vocabulary & Memory Work
- ☐ Vocabulary: star, nebulae, black hole
- ☐ Memory Work – Continue to work on memorizing the Types of Stars.

Sketch Assignment: Life Cycle of a Star
- Label the following: stellar nebulae, average star, red giant, planetary nebula, neutron star, the material condenses and begins to burn by converting hydrogen to helium, helium starts to convert to carbon and the star gets larger, the fuel in the core is used up and the outer layers are pushed out into space, the core shrinks to the size of a small planet

Writing Assignment
- Reading Assignment: *Kingfisher Science Encyclopedia* pg. 392-393 Stars
- Additional Research Readings:
  - Stars: *USE* pg. 158-159
  - The birth and death of stars: *DK Astro* pg. 60-61

Dates to Enter
- 1824-1910 – British astronomer William Huggins lives. He is the first to use spectroscopy in astronomy.
- 1906 – Ejnar Hertzsprung, a Danish astronomer, discovers that a star's temperature and luminosity are linked. He then arranges them into families, from hot and bright to cool and dim.

Sketch Assignment Lesson 3

Experiment: Why do stars twinkle?

Introduction

Stars are the most numerous heavenly bodies in our universe. Our galaxy alone has about 200 billion of them. Stars are balls of hot, glowing gas that produce light that we can see from Earth. Stars vary in size, but all appear very small from the Earth because they are so far away. At times stars appear to sparkle in the night sky. In this experiment, you are going to figure out why they appear to twinkle.

Hypothesis

I think that stars twinkle because ______________________________

____

____

____

Materials

______________ ______________

______________ ______________

______________ ______________

______________ ______________

______________ ______________

Procedure

____

____

____

____

____

____

____

____

Observations and Results

Conclusion

Written Assignment Lesson 3

Discussion Questions

1. What does light coming from a star tell an astronomer about a star?
2. Name three types of energy that are produced by a star.
3. How does a star's mass affect it?
4. Name two characteristics that are common to all stars.
5. What are the two types of star clusters? Explain both.
6. What happens to the core of a star when it dies?

Written Assignment Lesson 3

Student Assignment Sheet Astronomy Lesson 4
The Big and Little Dipper

Experiment: Shoebox Planetarium

Materials
- ✓ Shoebox
- ✓ Black construction paper
- ✓ Pin
- ✓ Flashlight

Procedure
1. Read the introduction on your experiment sheet.
2. At one end of the shoebox, cut out a hole that is a little less than 3 inches wide and 5 inches long.
3. Cut out two rectangles from the black construction paper measuring 3 inches by 5 inches. Then using the templates from your experiment sheet and a pin, punch out the necessary holes to make the Big Dipper and the Little Dipper constellations.
4. Tape one of the cards in front of the hole that was cut in the shoebox earlier. Place the flashlight in the box and cover with the lid. Turn off the lights in the room, angle the shoebox towards a wall, and observe what you see.
5. Repeat with the second card.

Vocabulary & Memory Work
- ☐ Vocabulary: constellation, planetarium
- ☐ Memory Work – Begin to work on memorizing the Constellations of the Zodiac: Aquarius, Aries, Cancer, Capricorn, Gemini, Leo, Libra, Pisces, Sagittarius, Scorpio, Taurus, Virgo

Sketch Assignment: Big Dipper & Little Dipper
- Label the following: Ursa Major, Ursa Minor, Polaris, Dubhe, Merak, Mizar, Benetnasch

Writing Assignment
- Reading Assignment: "Ursa Major and Ursa Minor," "Callisto and Arcas," and "The Two Bears" from *The Stars and their Stories* (found in the Appendix of the SG pp. 328-331)
- Additional Research Readings
  - None this week

Dates to Enter
- 2nd century – Ptolemy names forty-eight different constellations in his book *Almagest.*
- 17th-18th centuries – Forty more constellations are named, for a total of eighty-eight named constellations.

Sketch Assignment Lesson 4

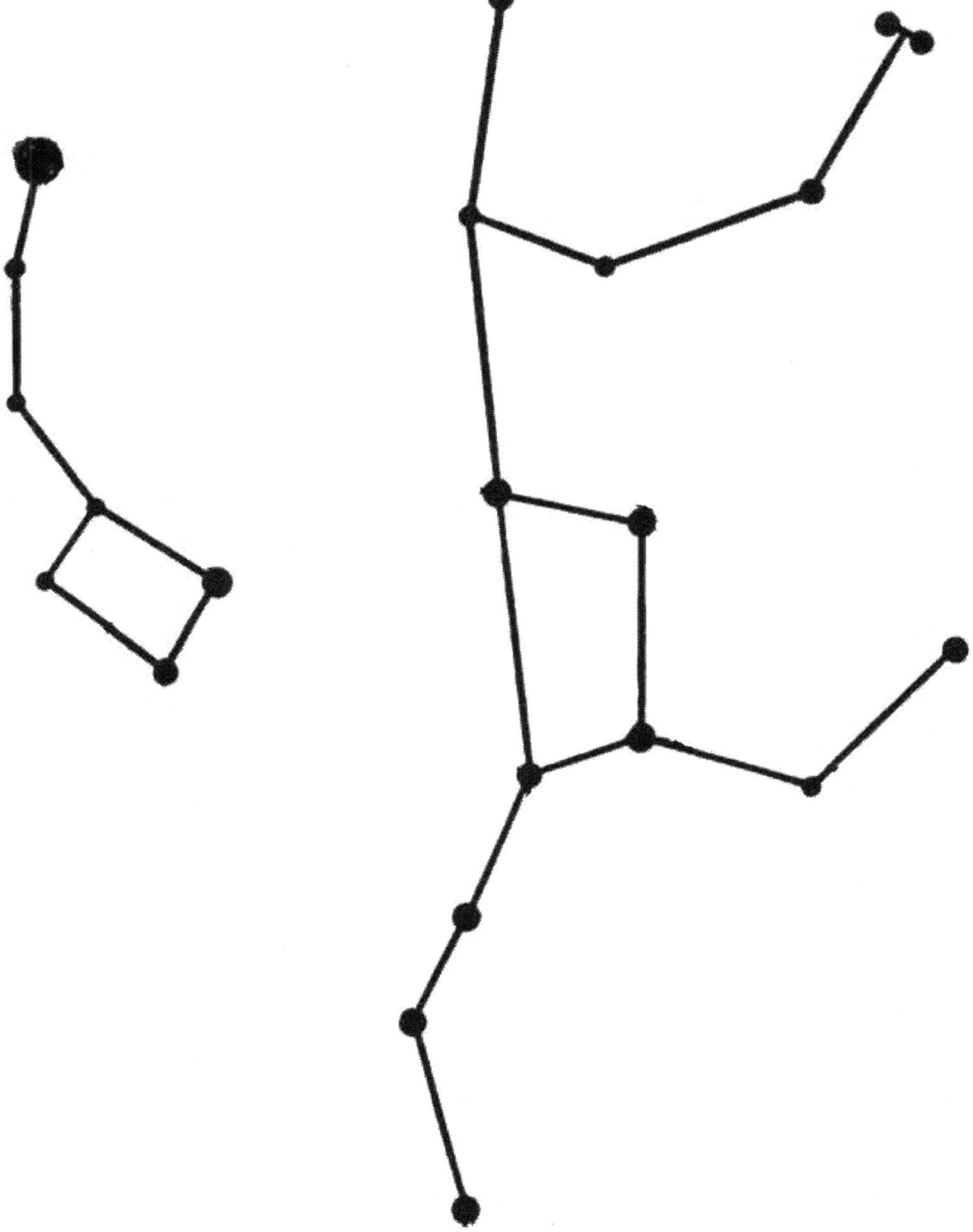

Experiment: Shoebox Planetarium

Introduction

A constellation is a pattern or shape that the group of stars makes in the sky. There are 88 named constellations in our sky. Your position on the Earth and the time of year determine which constellations you can see. The most recognized constellations in the Northern Hemisphere are the Big Dipper and the Little Dipper. In this experiment, you are going to make your own planetarium, or device used to project images of the stars, so that you can view these constellations at anytime.

Templates for the Planetarium Cards

The Big Dipper Template

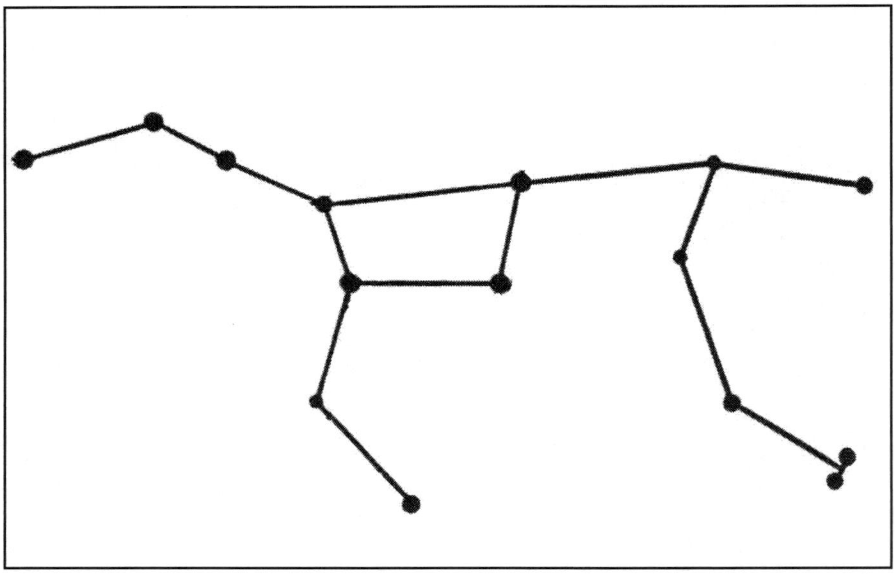

The Little Dipper Template

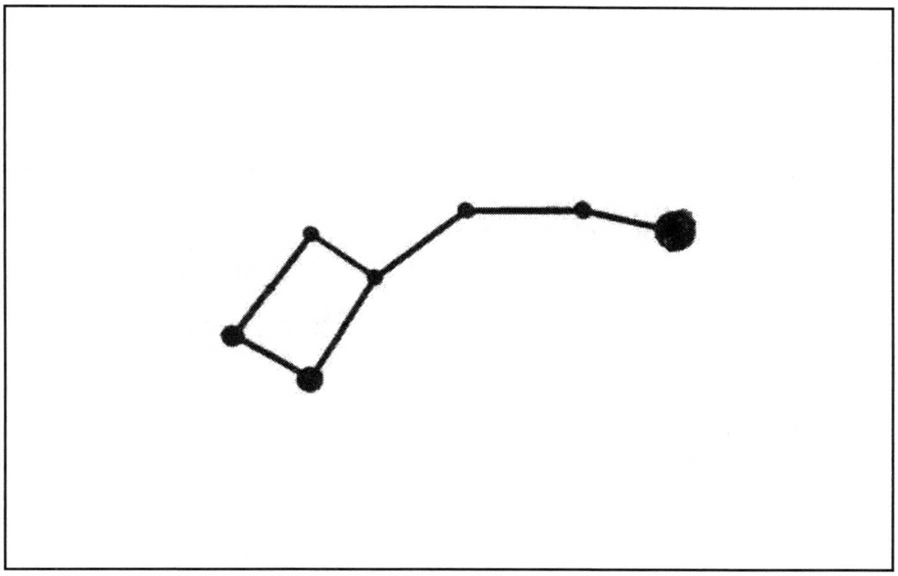

Written Assignment Lesson 4

Discussion Questions

1. Why are the constellations Ursa Major and Ursa Minor called the Big Dipper and Little Dipper?
2. Tell the story of the two bears in your own words.
3. What are some other names for the constellation Ursa Major?

Written Assignment Lesson 4

Student Assignment Sheet Astronomy Lesson 5
Constellations

Experiment: Constellations Research Project

This week you will spend part of the week on researching a constellation of the Zodiac and creating a profile page for that constellation.

Steps to Follow

1. Choose a Constellation – Choose one of the constellations of the Zodiac for an in-depth profile.
2. Do some research about the constellation – Use the internet and the resources you have in your home or at your library to find out more about your chosen constellation.
   Answer the following questions from your research:
   - ✓ What is the English translation of the Latin name of the constellation?
   - ✓ Where is the constellation found?
   - ✓ What are the major stars found in the constellation?
   - ✓ What is the best season to view the constellation?
   - ✓ What is the story behind the constellation's name?
   - ✓ Write down any interesting facts you have learned on individual index cards.
3. Complete the profile page for your constellation – Write the Latin name for your constellation on the blank at the top of the page. Then, fill out the remaining information. Your summary of the story behind the constellation should be four to five sentences long. Finally, draw your constellation in the box.

Writing Assignment

- Reading Assignment: *Kingfisher Science Encyclopedia* pg. 396-397 Constellations
- Additional Research Readings:
  - 🕮 Constellations: *USE* pg. 160
  - 🕮 Astrology: *DK Astro* pg. 16-17

Vocabulary & Memory Work

- ☐ Vocabulary: zodiac
- ☐ Memory Work – Continue to work on memorizing the Constellations of the Zodiac

Sketch Assignment: Constellations

- Label the following constellations and tell whether the are found in the Northern or Southern Hemisphere: The Southern Cross, Libra, Scorpio, Leo, Pegasus

Dates to Enter

- 🕒 7th century BC – Babylonian astronomers use a coordinate system resembling the Zodiac.
- 🕒 ca. 50 BC – A relief called the Dendera zodiac is the first known depiction of the classical zodiac of twelve signs.

Sketch Assignment Lesson 5

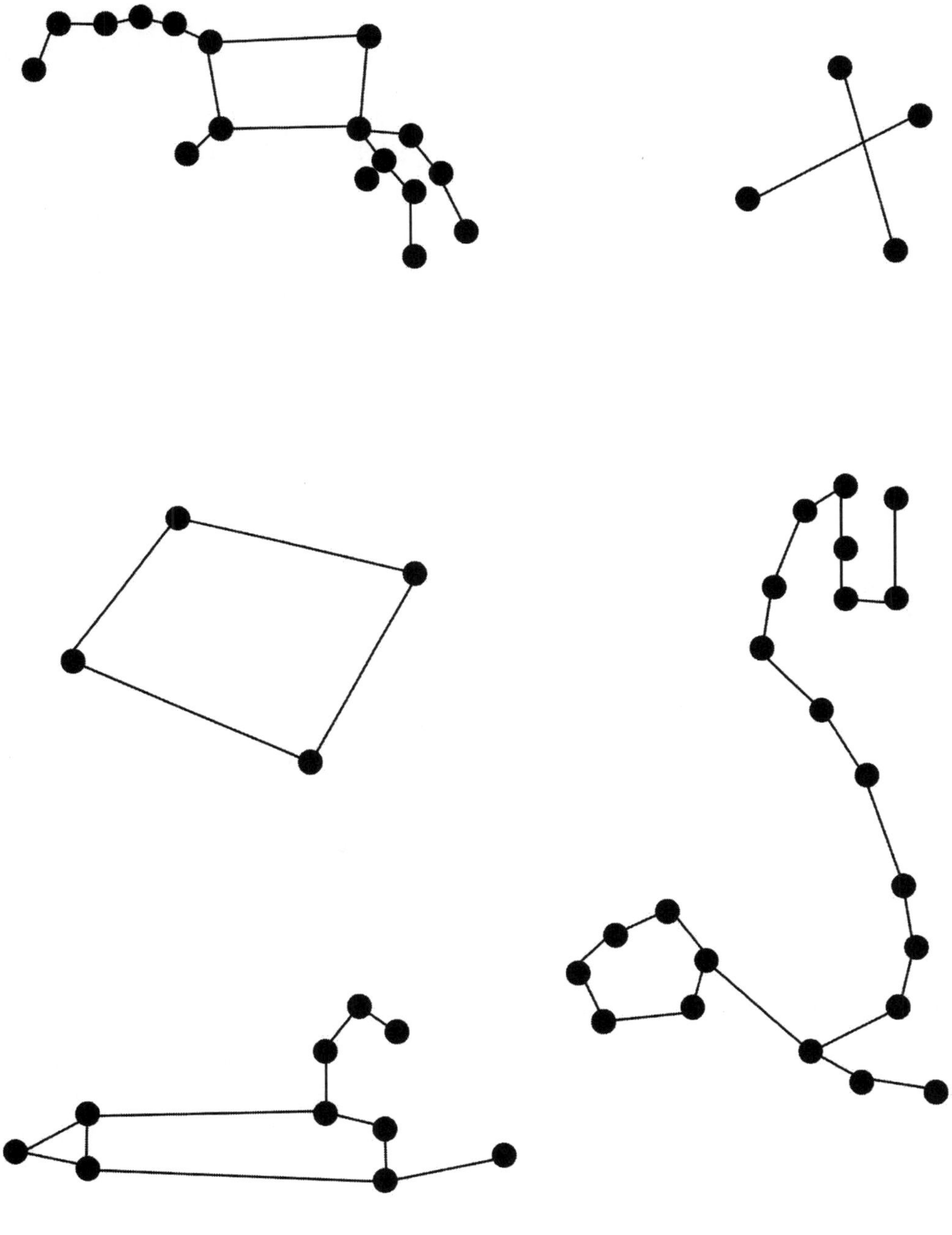

Constellation Profile Page

English Translation

Location

Major Stars

Best season to view the constellation

Story Behind the Name

Unit 2

Our Solar System

Unit 2: Our Solar System
Vocabulary Sheet

Define the following terms as they are assigned on your Student Assignment Sheet.

1. Photosphere – ______________________________

2. Prominence – ______________________________

3. Solar Wind – ______________________________

4. Sunspot – ______________________________

5. Crater – ______________________________

6. Planet – ______________________________

7. Axis – ______________________________

8. Orbit – ______________________________

9. Moon – ______________________________

____

____

10. Eclipse – ______________________________

____

____

11. Galilean moons – ______________________________

____

____

12. Gas giant – ______________________________

____

____

13. Dwarf planet – ______________________________

____

____

14. Asteroid – ______________________________

____

____

15. Comet – ______________________________

____

____

16. Meteor – ______________________________

____

____

17. Meteoroid – ______________________________

____

____

Student Assignment Sheet Astronomy Lesson 1
Sun

Experiment: Do the sun's rays contain heat?

Materials
- ✓ 2 glass jars or 2 clear glasses
- ✓ Plastic wrap
- ✓ 2 black tea bags
- ✓ Water
- ✓ Instant read thermometer

Procedure

****NOTE** – You will need to do this experiment on a sunny day.**

1. Read the introduction to this experiment and answer the question.
2. Fill each jar with the same amount of water. Take the temperature of the water in each jar. Then, place a tea bag in each jar and cover with the plastic wrap.
3. Set one of the jars outside in a sunny spot and leave the other jar inside your house.
4. After two hours, bring the jar you left outside into the house. Then, measure the temperature of the water and observe any changes to the appearance of the water. Write your results on your experiment sheet.
5. Draw conclusions and complete your experiment sheet.

Vocabulary & Memory Work

- ☐ Vocabulary: photosphere, prominence, solar wind, sunspot
- ☐ Memory Work – Begin to work on memorizing the Planet Order.

1. Sun
2. Mercury (Gravity: 0.38)
3. Venus (Gravity: 0.90)
4. Earth (Gravity: 1 or 9.8 m/s2) [Moon (Gravity: 0.17)]
5. Mars (Gravity: 0.38)
6. Jupiter (Gravity: 2.34)
7. Saturn (Gravity: 0.93)
8. Uranus (Gravity: 0.90)
9. Neptune (Gravity: 1.13)

Sketch Assignment: Parts of the Sun

- Label the Following: chromosphere, flare, facula, prominence, sunspot, hydrogen core, photosphere

Writing Assignment

- Reading Assignment: *Kingfisher Science Encyclopedia* pp. 394-395 The Sun
- Additional Research Readings
  - The Sun: *USE* pp. 192-193
  - The Sun: *DK Astro* pp. 38-39

Dates to Enter:

- 1930 – French astronomer, Bernard Lyot, invents the coronagraph, allowing scientists to view the Sun without waiting for a total solar eclipse.
- 1868-1938 – George Hale lived. He was an American astronomer who studied sunspots and discovered the magnetic fields within them.

Sketch Assignment Lesson 1

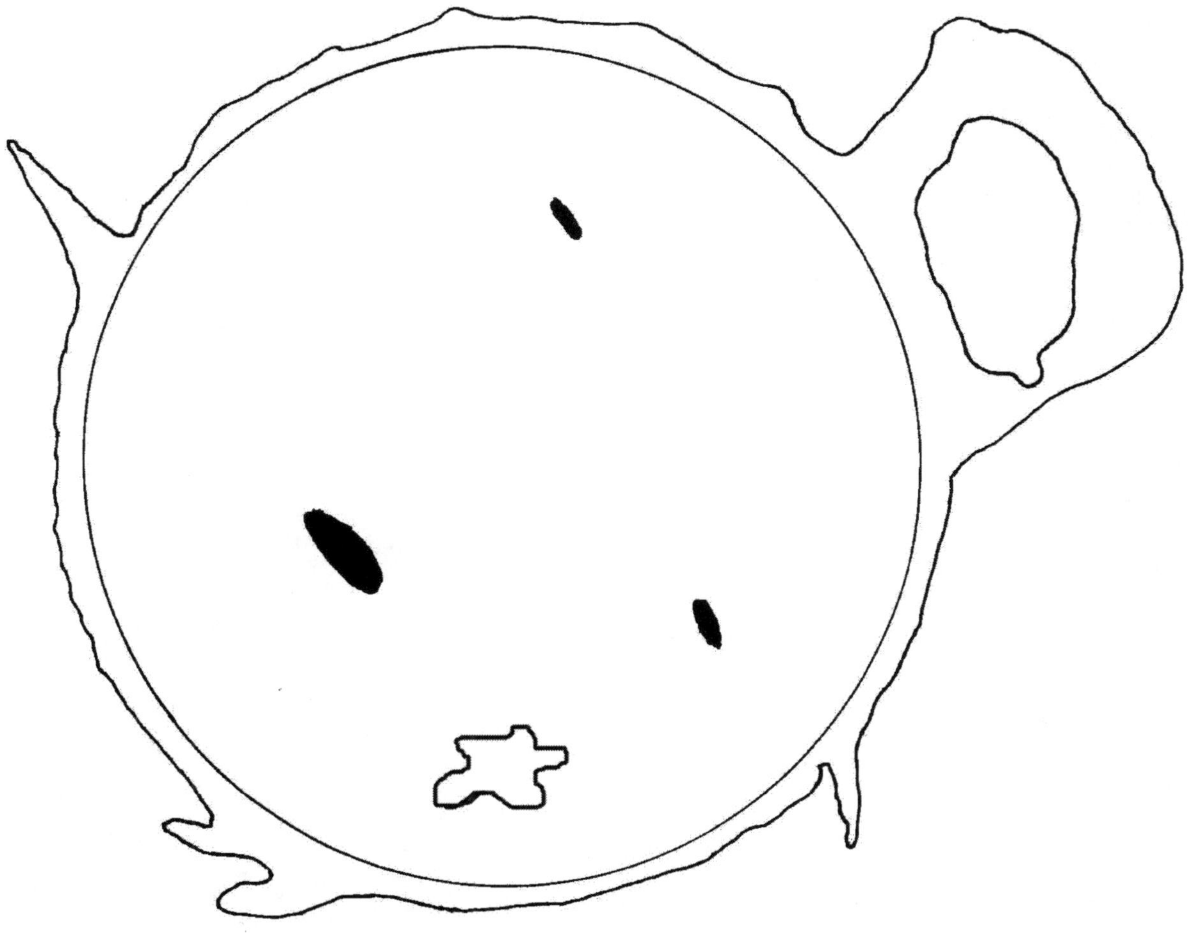

Experiment: Do the sun's rays contain heat?

Introduction

The sun is the closest star to the earth. It is a giant ball of a constantly exploding gas called hydrogen. The surface of the sun is around 9,932°F, but the core is much hotter. It is the largest object in our Solar System. The sun's gravitational pull is so strong that all the planets in our Solar System revolve around it. The sun is important to the earth in many ways. In this experiment, you are going to look specifically at the sun's rays affect the objects on our planet.

Hypothesis

I think that the Sun's rays (do / do not) contain heat.

Materials

________________ ________________

________________ ________________

________________ ________________

________________ ________________

________________ ________________

Procedure

____

____

____

____

____

____

____

____

____

____

____

Observations and Results

Temperature of the Jars

	Initial Temperature	Final Temperature
Outside Jar		
Inside Jar		

*Be sure to include units for your temperature measurements.

Appearance of the Jars

Outside Jar	Inside Jar

Conclusion

Written Assignment Lesson 1

Discussion Questions

1. What is the sun?
2. What causes sunspots?
3. What is the chromosphere of the sun composed of?
4. Briefly explain the expected life of the sun.

Written Assignment Lesson 1

Student Assignment Sheet Astronomy Lesson 2
Inner Planets-Mercury, Venus, and Mars

Experiment: Can yeast survive on Mercury or Venus?

> **CAUTION**
> ***Hot water is very dangerous and can cause injury. Be sure to have an adult handle the hot water at all times. ALWAYS use proper protection when handling hot water!***

Materials
- ✓ 3 heavy duty plastic bottles
- ✓ 3 balloons
- ✓ 3 tsp of yeast (1 ½ packages)
- ✓ 3 TBSP of sugar
- ✓ Water
- ✓ White vinegar

Procedure
1. Read the introduction to this experiment and answer the question.
2. Label the bottles, "Mercury," "Venus," and "Earth." Use a funnel to carefully add 1 ½ cups of warm (between 125°F and 130°F) water to the Earth bottle. Use a funnel to carefully add 1½ cups of ice cold (below 40°F) water to the Mercury bottle. Have an adult use a funnel to carefully add 1 cup of hot (above 140°F) water and ½ cup of vinegar to the Venus bottle.
3. Add 1 tsp of yeast and one TBSP of sugar to each of the bottles, and then quickly cover the top of the bottle with a balloon.
4. Wait 30 minutes. Then, make observations and measure the circumference of each balloon.
5. Draw conclusions and complete your experiment sheet.

Vocabulary & Memory Work
- ☐ Vocabulary: crater, planet
- ☐ Memory Work – Continue to work on memorizing on Planet Order

Sketch Assignment: Inner Planets
- Label the Following: Mercury, Caloris Basin, crater-covered surface; Venus, multiple layers of dense clouds; Mars, Polar regions covered with water, ice and carbon dioxide ice

Writing Assignment
- Reading Assignment: *Kingfisher Science Encyclopedia* pg. 403 Mercury & pg. 404 Venus, pg. 405 Mars
- Additional Research Readings
  - The Inner Planets: *USE* pg. 164-165
  - Mercury: *DK Astro* pg. 44-45, Venus: *DK Astro* pg. 46-47, Mars: *DK Astro* pg. 48-49
  - Venus & Mercury: *ENS* pg. 28-29, *ENS* pg. 30-31

Dates to Enter
- 1974-1975 – *Mariner 10* flies past Mercury three times to take overlapping photos of the surface of the planet.
- 1990-1994 – *Magellan* maps 98% of the surface of Venus.
- 1976 – Two American space probes land on Mars to test for signs of life, but find none.
- 1997 – *Pathfinder* lands on Mars and delivers the robotic rover *Sojourner* to explore the surface of Mars.
- 2004 – Mobile robots, called *Spirit* and *Opportunity,* explore the surface of Mars.

Sketch Assignment Lesson 2

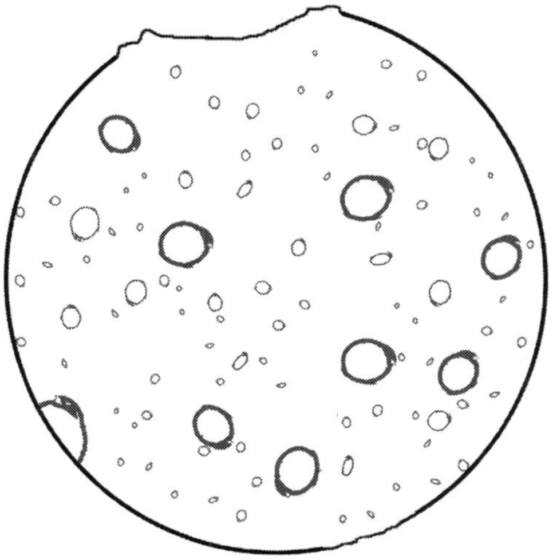

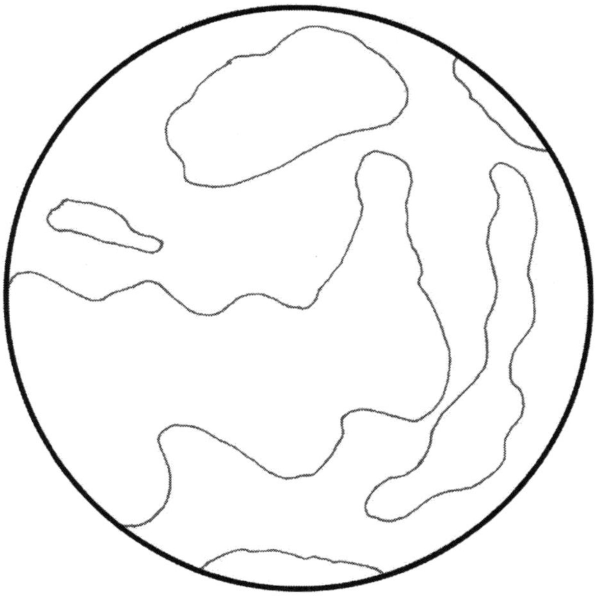

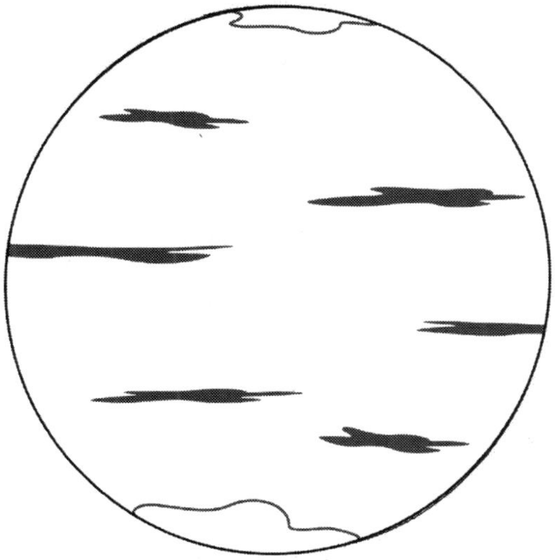

Experiment: Can yeast survive on Mercury or Venus?

Introduction

Each planet has its own unique atmosphere or layer of gas surrounding the planet. The atmosphere and the planet's proximity to the Sun affect the surface conditions and temperature of the sphere. The atmosphere around Earth is perfectly suited to support life on the planet. In this experiment, you are going to test and see if Mercury and Venus could also support life.

Hypothesis

Can life survive on Mercury? yes no

Can life survive on Venus? yes no

Materials

_______________ _______________

_______________ _______________

_______________ _______________

_______________ _______________

_______________ _______________

Procedure

____

____

____

____

____

____

____

____

____

Observations and Results

Bottle	Circumference of the Balloon
Mercury	
Venus	
Earth	

Conclusion

Written Assignment Lesson 2

Discussion Questions

Mercury pg. 403

1. What is the surface of Mercury like?
2. What two factors contribute to the fact that a day on Mercury is longer than a year?

Venus pg. 404

1. Why does Venus appear as a bright star in our sky?
2. What are two reasons why humans could not survive on Venus?
3. What is the surface of Venus like?

Mars pg. 405

1. Why is Mars known as the Red Planet?
2. Has life been found on Mars? What have explorers found?
3. What is the surface of Mars like?

Written Assignment Lesson 2

Student Assignment Sheet Astronomy Lesson 3
Earth and the Moon

Experiment: Does the moon change size?

Materials
- ✓ Apple
- ✓ Fork
- ✓ Flashlight
- ✓ Partner

Procedure
1. Read the introduction to this experiment and answer the question.
2. Secure the apple on the end of the fork.
3. Have your partner stand 5 feet away from you and shine the flashlight in your direction.
4. Hold the fork with the apple at arm's length in front of the flashlight and a little above the beam. ***(Note** – The apple should appear darkened.)*
5. Keep the apple in the same place and turn slowly counterclockwise around an imaginary circle with the apple as the center. Continue until you have completed your circle and are back to your starting point. ***(Note** – You may need to duck down a bit as you turn so that you don't block the flashlight beam with your body.)*
6. Write down your observations about what happened to your apple as you were turning.
7. Draw conclusions and complete your experiment sheet.

Vocabulary & Memory Work

- ☐ Vocabulary: axis, orbit, moon, eclipse
- ☐ Memory Work – Continue to work on memorizing the Planet Order

Sketch Assignment: Phases of the Moon

- Label the following and color in the portion of the moon that the labels represent: New Moon, Waxing Crescent, First Quarter, Gibbous Waxing, Full Moon, Gibbous Waning, Last Quarter, and Waning Crescent

Writing Assignment

- Reading Assignment: *Kingfisher Science Encyclopedia* pg. 400-401 Earth and the Moon, pg. 402 Eclipses
- Additional Research Readings
  - The Earth and the Moon: *USE* pg. 166-167
  - Earth: *DK Astro* pg. 42-43
  - Moon: *DK Astro* pg. 40-41

Dates to Enter

- 1647 – Johannes Hevelius publishes the first lunar atlas.
- October 1959 – Russian spacecraft *Luna 3* transmits the first images of the far side of the moon back to the Earth.
- July 20, 1969 – US astronauts Neil Armstrong and Edwin Aldrin became the first humans to walk on the moon.

Sketch Assignment Lesson 3

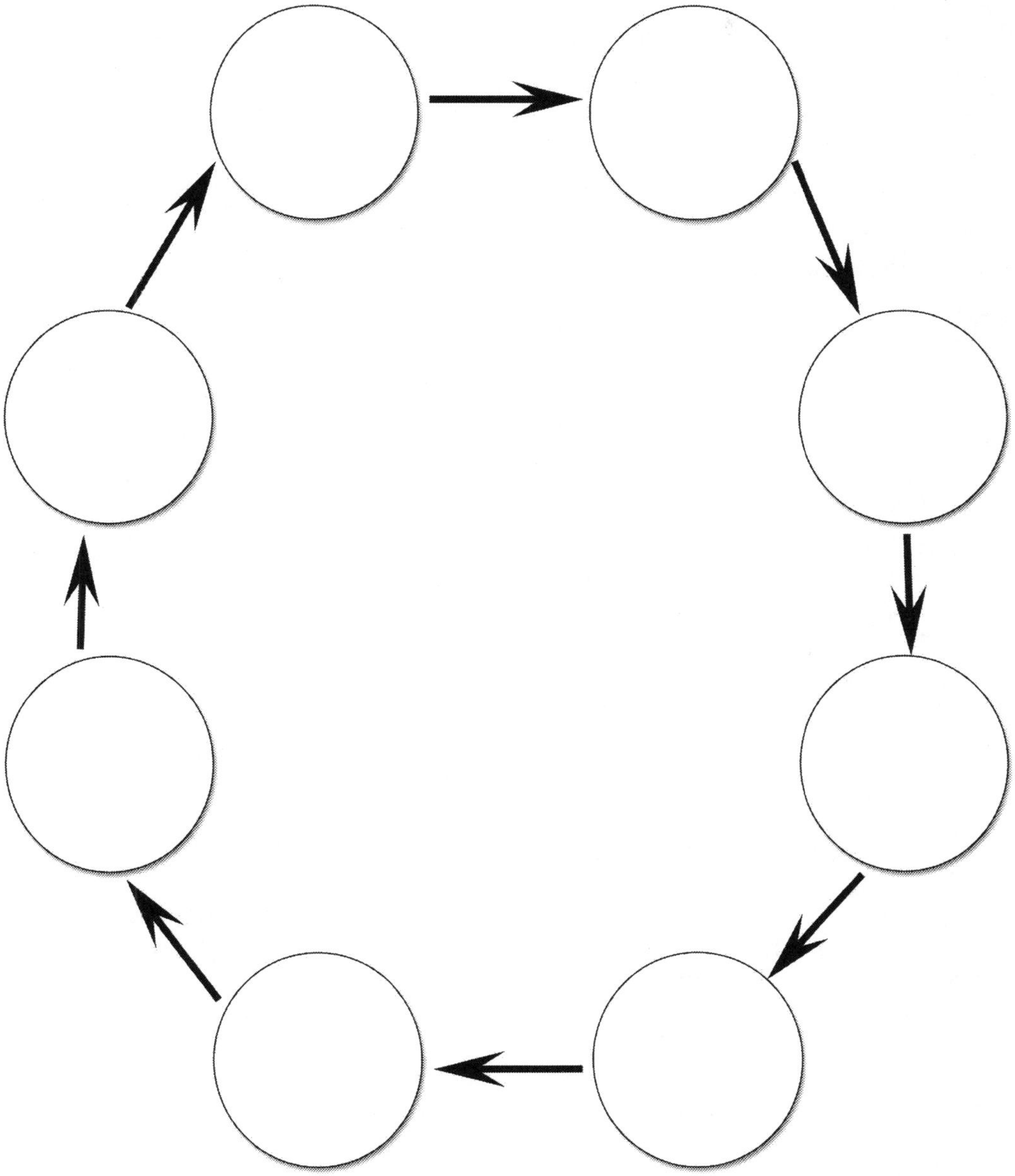

Experiment: Does the moon change size?

Introduction

The moon is a ball of rock that orbits around the Earth. It takes 27.3 days to complete its orbit around our planet. As the moon travels around the Earth it appears to be smaller or larger to us. In this experiment, you are going to determine if the moon actually changes size as it orbits around us by using an apple for the moon, a flashlight for the sun and your head for the Earth.

Hypothesis

Does the moon change size? yes no

Materials

______________________ ______________________

______________________ ______________________

______________________ ______________________

______________________ ______________________

______________________ ______________________

Procedure

____

____

____

____

____

____

____

____

____

____

Observations and Results

Conclusion

Written Assignment Lesson 3

Discussion Questions

Earth and the Moon pg. 400-401

1. What makes the Earth unique to our solar system?
2. How long does it take for the Earth to orbit the sun one time? How long does it take the Earth to complete one spin on its axis?
3. Why do we only see one side of the moon from the Earth?
4. What is a moon phase?

Eclipses pg. 402

1. What is the difference between a solar eclipse and a lunar eclipse?

Written Assignment Lesson 3

Student Assignment Sheet Astronomy Lesson 4
Outer Planets-Jupiter and Saturn

Experiment: Does the heat in Jupiter's core affect the storms on its surface?

Materials

- ✓ Wooden dowel rod
- ✓ Paper clip
- ✓ 4 inch square of aluminum foil (not heavy duty)
- ✓ Toaster

CAUTION: ***Be careful not to touch your toaster or your foil kite until they have time to cool. They will be very hot and can burn you.***

Procedure

1. Read the introduction to this experiment and answer the question.
2. Begin by making a kite. First, unbend your paper clip halfway. Place the remaining loop of the paper clip over your wooden dowel (it should be a tight fit.) Then, puncture the corner of your aluminum foil with the straight end of the paper clip. Finally, bend the straight end of the paper clip over so that the foil is secure.
3. Turn your toaster on and wait 1 minute. Move your foil kite so that it is about a foot above the toaster's opening. Observe what happens to your kite and write it down on your experiment sheet.
4. Remove your kite from over the heat and turn off the toaster. Observe what happens to your kite and write it down on your experiment sheet.
5. Draw conclusions and complete your experiment sheet.

Vocabulary & Memory Work

- ☐ Vocabulary: Galilean moons, gas giants
- ☐ Memory Work – Continue to work on memorizing the Planet Order

Sketch Assignment: Jupiter & Saturn

- Label the following: Jupiter, the Red Spot, bands of storms, Saturn, rings, bands of ammonia, water and methane clouds

Writing Assignment

- Reading Assignment: *Kingfisher Science Encyclopedia* pg. 406 Jupiter, pg. 407 Saturn
- Additional Research Readings
  - The Outer Planets: *USE* pg. 168-169
  - Jupiter: *DK Astro* pg. 50-51, *ENS* pg. 32-33
  - Saturn: *DK Astro* pg. 52-53, *ENS* pg. 34-35

Dates to Enter

- 1610 – Galileo makes the first systematic study of Jupiter's 4 largest moons.
- 1660 – Robert Hooke reports a giant spot on Jupiter's surface.
- 1655 – Dutch scientist, Christiaan Huygens, correctly identifies Saturn's rings.
- 1979 – *Pioneer II* explores Saturn's rings.
- 1995 – The *Galileo* probe reaches Jupiter. It studies Jupiter's atmosphere and its moons.
- 2004 – *Cassini* arrives at Saturn to study the planet's moons and rings.

Sketch Assignment Lesson 4

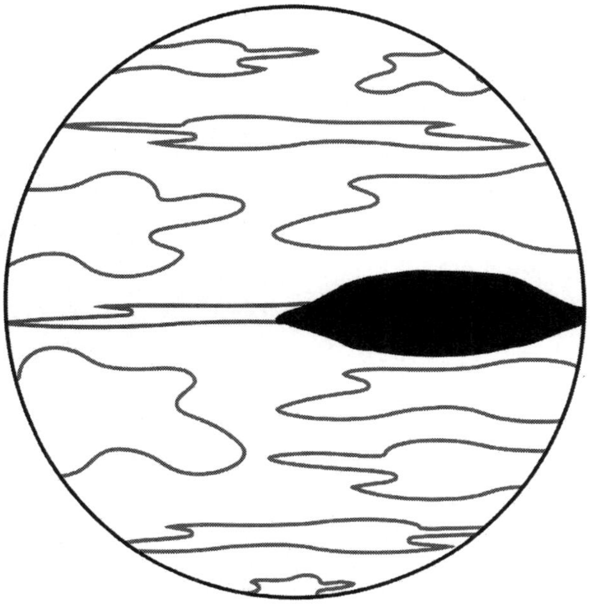

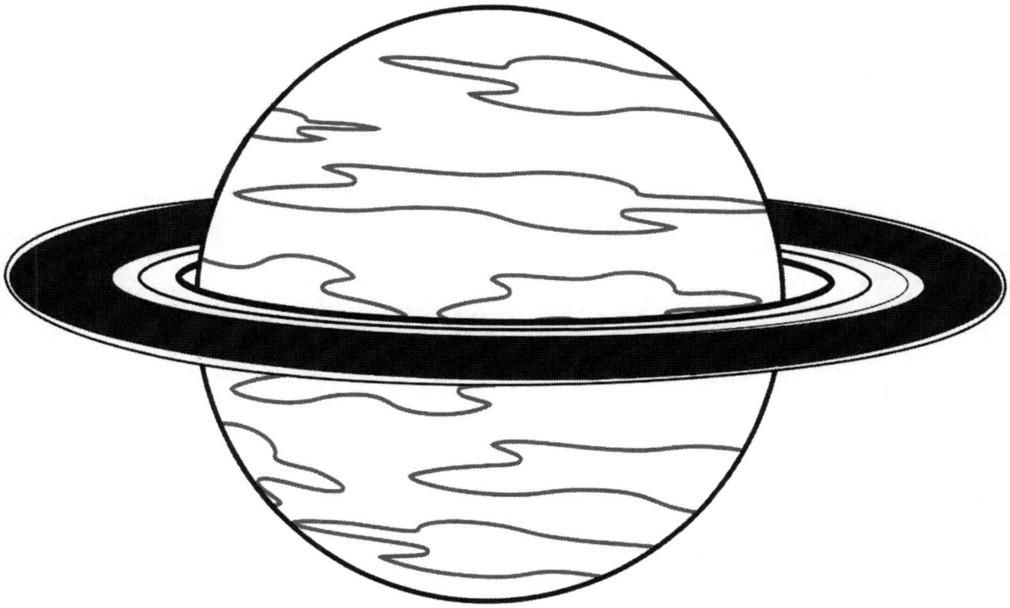

Experiment: Does the heat in Jupiter's core affect the storms on its surface?

Introduction

Jupiter is a very turbulent planet. Its many bands and prominent red spot are evidence of a stormy atmosphere that surrounds the planet. Jupiter's Red Spot is known as the solar system's oldest hurricane. The high pressure that is found on the planet contributes to the perpetuation of these storms, but there are other factors involved as well. In this experiment, you will examine whether the heat from Jupiter's core is able to create winds that help to perpetuate the storms that cover the planet.

Hypothesis

I think that the heat in Jupiter's core (does / does not) affect the storms on its surface.

Materials

______________________ ______________________

______________________ ______________________

______________________ ______________________

______________________ ______________________

______________________ ______________________

Procedure

____

____

____

____

____

____

____

____

Observations and Results

Conclusion

Written Assignment Lesson 4

Discussion Questions

Jupiter pg. 406

1. Describe what Jupiter is like.
2. What are two factors that contribute to Jupiter's stormy atmosphere?

Saturn pg. 407

1. What is the outer surface of Saturn like?
2. What are Saturn's rings made from? Where are the largest pieces of Saturn's rings found?

Written Assignment Lesson 4

Student Assignment Sheet Astronomy Lesson 5
Outer Planets-Uranus, Neptune, and Minor Members

Experiment: Scaled Model of the Solar System

Materials
- ✓ Construction paper
- ✓ Paints or crayons

Procedure
1. This week you will be making a scaled version of the solar system, using the chart on your experiment sheet for your placements. Try to make your planets look as realistic as possible by painting or drawing the major features onto your models. For the Sun, tape four pieces of construction paper together lengthwise and then trim them up to look like a portion of a circle. Add solar flares, a prominence, sunspots, and facula.

	Distance from the Sun	Scale diameter of planet
Mercury	2 inches	3/4 inch
Venus	3 inches	1 3/4 inches
Earth	4 inches	2 inches
Mars	6 inches	1 1/8 inches
Jupiter	1 foot, 9 inches	22 inches
Saturn	3 feet, 2 inches	20 inches
Uranus	6 feet, 5 inches	8 inches
Neptune	10 feet, 1 inch	7 1/2 inches
Pluto (Optional)	13 feet, 3 inches	1/2 inches

2. Take a picture of your model and write what you have learned about the solar system from making the model on your experiment sheet.

Vocabulary & Memory Work
- ☐ Vocabulary: dwarf planet, asteroid
- ☐ Memory Work – Continue to work on memorizing the Planet Order

Writing Assignment
- Reading Assignment: *Kingfisher Science Encyclopedia* pg. 408 Uranus, pg. 409 Neptune and pg. 410-411 The Solar System's Minor Members
- Additional Research Readings:
  - 📖 The Outer Planets: *USE* pg. 170-171
  - 📖 Uranus: *DK Astro* pg. 54-55
  - 📖 Neptune & Pluto: *DK Astro* pg. 56-57

Dates to Enter
- March 13, 1781 – William Herschel discovers Uranus using a homemade telescope.
- 1846 – Johann Galle finds the planet Neptune.
- 1930 – Clyde Tombaugh, a US astronomer, discovers Pluto.
- 2006 – Pluto is reclassified as a dwarf planet.

Scaled Model of the Solar System

What I learned from this project

Written Assignment Lesson 5

Discussion Questions

Uranus pg. 408

1. What are some key features of Uranus?

Neptune pg. 409

1. What are some key features of Neptune?
2. How was Neptune discovered?

The Solar System's Minor Members pg. 410-411

1. What are the four main groups of minor members in our solar system?
2. Why does Pluto's distance from the Sun vary?
3. Why are the asteroids in the asteroid belt called the minor planets?

Written Assignment Lesson 5

Student Assignment Sheet Astronomy Lesson 6
Comets and Meteors

Experiment: Does the size of a meteor change the impact it will have on a planet's surface?

Materials

- ✓ Marble
- ✓ Rubber bouncy ball
- ✓ Tennis ball
- ✓ Cake pan
- ✓ Cornstarch
- ✓ Ruler
- ✓ Foam ball (a little larger than the tennis ball)

Procedure

1. Read the introduction to this experiment and answer the question.
2. Pour a layer of cornstarch on the bottom of the cake pan about ½ inch deep. Shake lightly so that the surface is smooth.
3. Drop the marble from a height of 2 feet, aiming for the center of the pan. Observe what happens. Remove the marble, being careful not disturb the cornstarch, and measure the width and depth of the crater created. Record the measurement on your experiment sheet and shake the pan lightly so that the surface is smooth.
4. Repeat step #3 for the rubber bouncy ball, foam ball, and tennis ball.
5. Draw conclusions and complete your experiment sheet.

Vocabulary & Memory Work

- ☐ Vocabulary: comet, meteor, meteoroid
- ☐ Memory Work – Continue to work on memorizing the Planet Order

Sketch Assignment: Anatomy of a Comet

- Label the following: nucleus, dust tail, gas tail

Writing Assignment

- Reading Assignment: *Kingfisher Science Encyclopedia* pg. 412 Comets, pg. 413 Meteors and Meteorites
- Additional Research Readings
  - Space Debris: *USE* pg. 172-173
  - Travelers in Space: *DK Astro* pg. 58-59

Dates to Enter

- 1997 – The comet *Hale-Bopp* is in the Earth's night sky. It won't return for another 2,400 years.
- 1656-1742 – English astronomer, Edmond Halley lives. He correctly predicts that a comet will return to Earth's night sky in 1758, 1835, and 1910.
- 1976 – The second largest iron & nickel meteorite in the United States is found.
- 1986 – The space probe *Giotto*, is sent inside Halley's comet, giving people the first look at a comet's nucleus.

Sketch Assignment Lesson 6

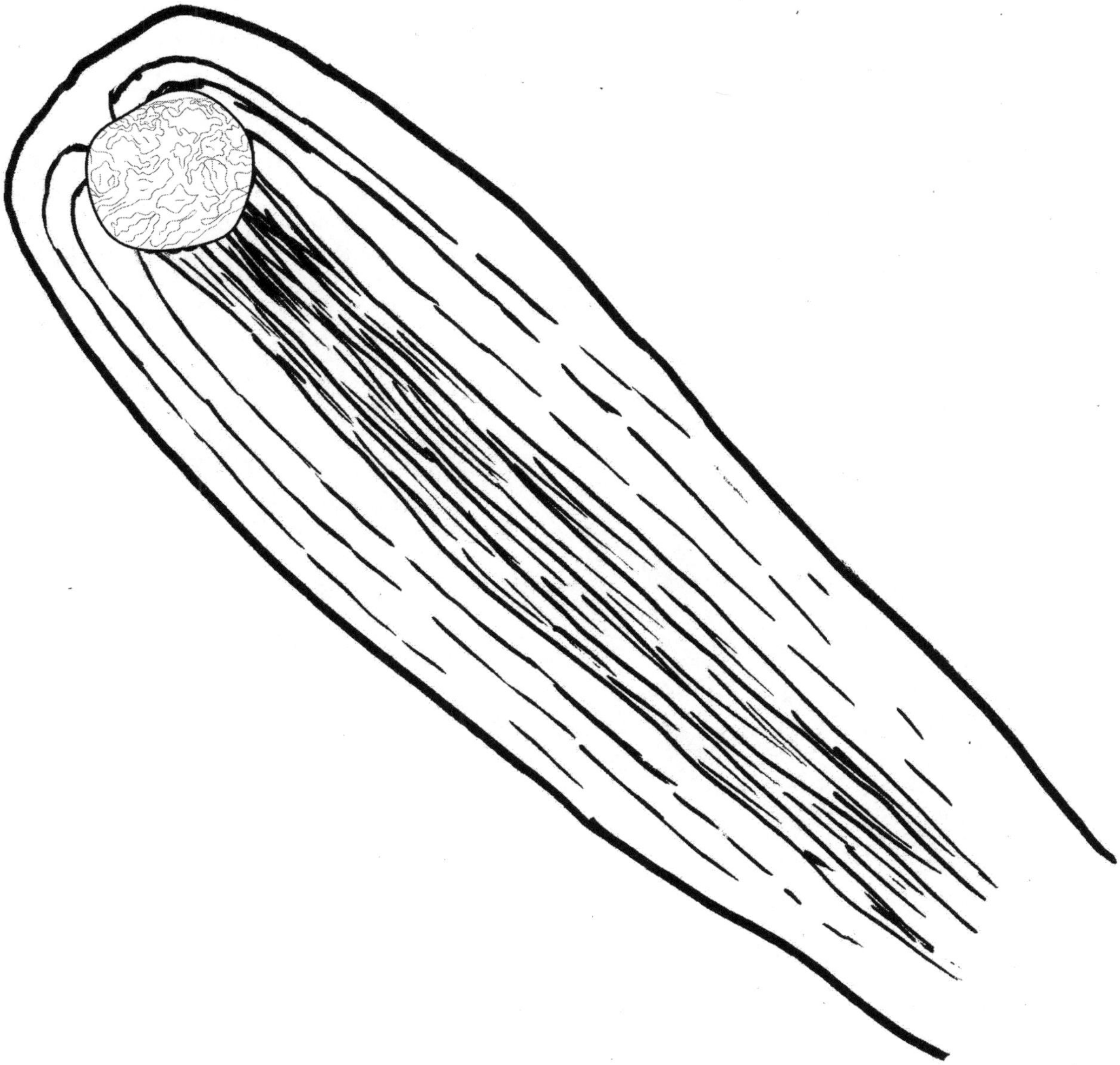

Experiment: Does the size of a meteor change the impact it will have on a planet's surface?

Introduction:

Bits of space dust and iron-rich rock, called meteoroids, can be found throughout the solar system. They are constantly bombarding the planets. On Earth, most of these meteors burn up before they hit the surface due to our atmosphere, but this is not always the case on other planets. The surfaces of those planets are dotted with impact craters from the meteors. In this experiment, you are going to test whether the size of the meteor affects the size of the crater it will make.

Hypothesis (What I think the answer to the question will be):

I think that the (larger / smaller) a meteor is the (larger / smaller)

the impact crater will be.

Materials (What we used):

______________________ ______________________

______________________ ______________________

______________________ ______________________

______________________ ______________________

______________________ ______________________

Procedure (What we did):

____

____

____

____

____

____

____

____

Observations & Results (What I saw):

	Width of Crater	Depth of Crater
Marble		
Rubber Bouncy Ball		
Tennis Ball		
Foam Ball		

*Be sure to include units for your measurements.

Conclusion (What I learned):

Written Assignment Lesson 6

Discussion Questions

Comet pg. 412

1. What is common to all comets?
2. Why do comets grow a tail when they are closer to the sun?

Meteors & Meteorites pg. 413

1. What is a shooting star?
2. What are meteorites?

Written Assignment Lesson 6

Unit 3

The Periodic Table

Unit 3: The Periodic Table

Vocabulary Sheet

Define the following terms as they are assigned on your Student Assignment Sheet.

1. Atom – ____

____

____

2. Electron Shell – ____

____

____

3. Element – ____

____

____

4. Compound – ____

____

____

5. Period – ____

____

____

6. Group – ____

____

____

7. Atomic Number – ____

____

____

8. Atomic Mass – ____

____

____

9. Metal – ____

10. Poor Metal –

11. Semimetal –

12. Malleable –

13. Semiconductor –

14. Nonmetal –

Student Assignment Sheet Lesson 1
Elements and Atoms

Research Report: Element Profile Page

This week you will spend time researching an element of your choice and creating a profile page for that element.

Steps to Follow:

1. Begin by choosing one of the elements of the periodic table for an in-depth profile.
2. Do some research to find out more about your chosen element. Use the internet and the resources you have in your home or at your library to find out more about the element.
3. Answer the following questions from your research:
   - ✓ What is the name, symbol, atomic number and atomic mass of the element?
   - ✓ Who discovered the element and when did that happen?
   - ✓ Is it a gas, liquid or solid at room temperature?
   - ✓ Where is the element typically found?
   - ✓ What are the major uses of the element?

   Be sure to also write down any additional interesting facts you have learned on individual index cards.
4. Finally, complete the profile page for your element. Fill out the top left-hand rectangle with the information on the element from the periodic table. Then, answer the questions and write a brief summary on the element and its uses. You may also want to include the story of how the element was discovered in your summary.

Vocabulary & Memory Work

- ☐ Vocabulary: atom, electron shell, element, compound
- ☐ Memory Work—This year you will be memorizing the elements that make up the periodic table, along with their atomic number and symbol. This week, work on the following elements:
  - ✓ 1-H-Hydrogen, 2-He-Helium, 3-Li-Lithium, 4-Be-Beryllium

Sketch: Parts of an Atom

- Label the following: electron orbit, nucleus, electron shell, and the electron, proton, and neutron at the bottom of the page

Writing

- Reading Assignment: *DK Encyclopedia of Science* pp. 24-25 (Atomic Structure) and pg. 31 (Elements)
- Additional Research Readings:
  - The Elements: *DK Chem* pp. 18-19
  - Atoms: *UDIS* pg. 124
  - The Elements: *KSE* pp. 148-149 and Atoms: *KSE* pp. 150-151

Dates

- 340 BC – Aristotle proposes that all substances are made up of combinations of four elements: earth, air, water and fire.
- 1766-1844 – John Dalton lives. He is responsible for writing the first atomic theory.
- 1897 – J.J. Thompson discovers the electron.
- 1909 – Rutherford, along with two other scientists, discovers the nucleus of an atom.
- 1913 – Niels Bohr comes up with the Bohr model of an atom.

Sketch Lesson 1

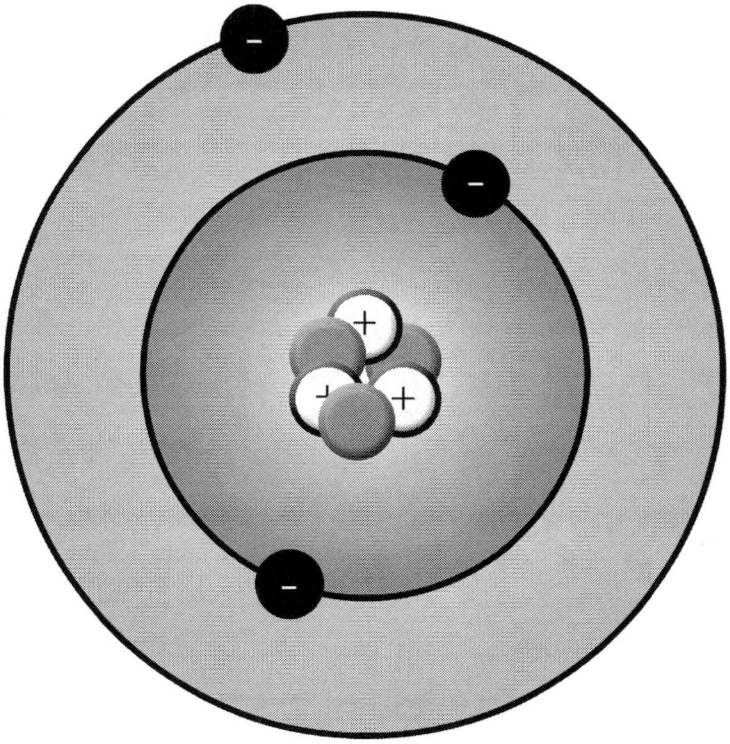

Element Profile Page

Discovered by: ____________________________

Discoverd in: ____________________________

Common uses: ____________________________

Typically found in: ____________________________

Interesting Facts: ____________________________

Written Assignment Lesson 1

Discussion Questions

pp. 24-25, Atomic Structure

1. What is an atom?
2. What is the basic structure of an atom?
3. What are the charges of the subatomic particles (i.e. protons, neutrons and electrons)?

pg. 31, Elements

1. How many of the known elements occur naturally on the Earth?
2. What are the two most common elements in the universe?

Written Assignment Lesson 1

Student Assignment Sheet Lesson 2
Periodic Table

Experiment: Making Ice Cream

Materials:

- ✓ Heavy cream
- ✓ Milk
- ✓ Sugar
- ✓ Vanilla
- ✓ Crushed ice
- ✓ 1 small & 1 large zip-locking plastic bag
- ✓ Dish towel or oven mitt
- ✓ Rock salt

Procedure:

1. Mix together the ½ cup cream (120 mL), 1 cup milk (240 mL), ½ cup sugar (225 g) and 1 tsp vanilla (5 mL) in the small zip-locking plastic bag. Then, add the crushed ice and rock salt to the large zip-locking plastic bag. (***Note:*** *Make sure that the air is removed from both bags and they are sealed tightly.*)
2. Place the smaller bag with the cream mixture inside the larger bag with the ice. Cover the outside of the bag with the dish towel (or put on the oven mitt) and massage or shake the bag until the cream mixture has frozen. It should take about 5 to 10 minutes.
3. Take the smaller bag out of the larger bag, wipe off the salt water, open, eat and enjoy!

Vocabulary & Memory Work

- ☐ Vocabulary: period, group, atomic number, atomic mass
- ☐ Memory Work—This week, add the following elements to what you are working on memorizing:
  - ✓ 5-B-Boron, 6-C-Carbon, 7-N-Nitrogen, 8-O-Oxygen

Sketch: The Periodic Table

- Label the following on the periodic table: periods 1-7, groups I-XVII, color the categories and fill in the categories key according to the table found on pg. 32-33 of the *DK Encyclopedia of Science*, with the exception of elements 104-118, which are known as Superheavy Elements
- Label the following on the element call out box: atomic number, atomic mass, symbol, element name

Writing

- Reading Assignment: *DK Encyclopedia of Science* pp. 32-33 (The Periodic Table)
- Additional Research Readings:
  - The Periodic Table: *KSE* pp. 152-153
  - The Periodic Table: *UDIS* pp. 164-165
  - The Periodic Table: *DK Chem* pp. 22-23

Dates

- 1869 – Russian chemist Dmitri Mendeleyev draws up the very first periodic table, leaving gaps for elements that had not yet been discovered.

Sketch Lesson 2

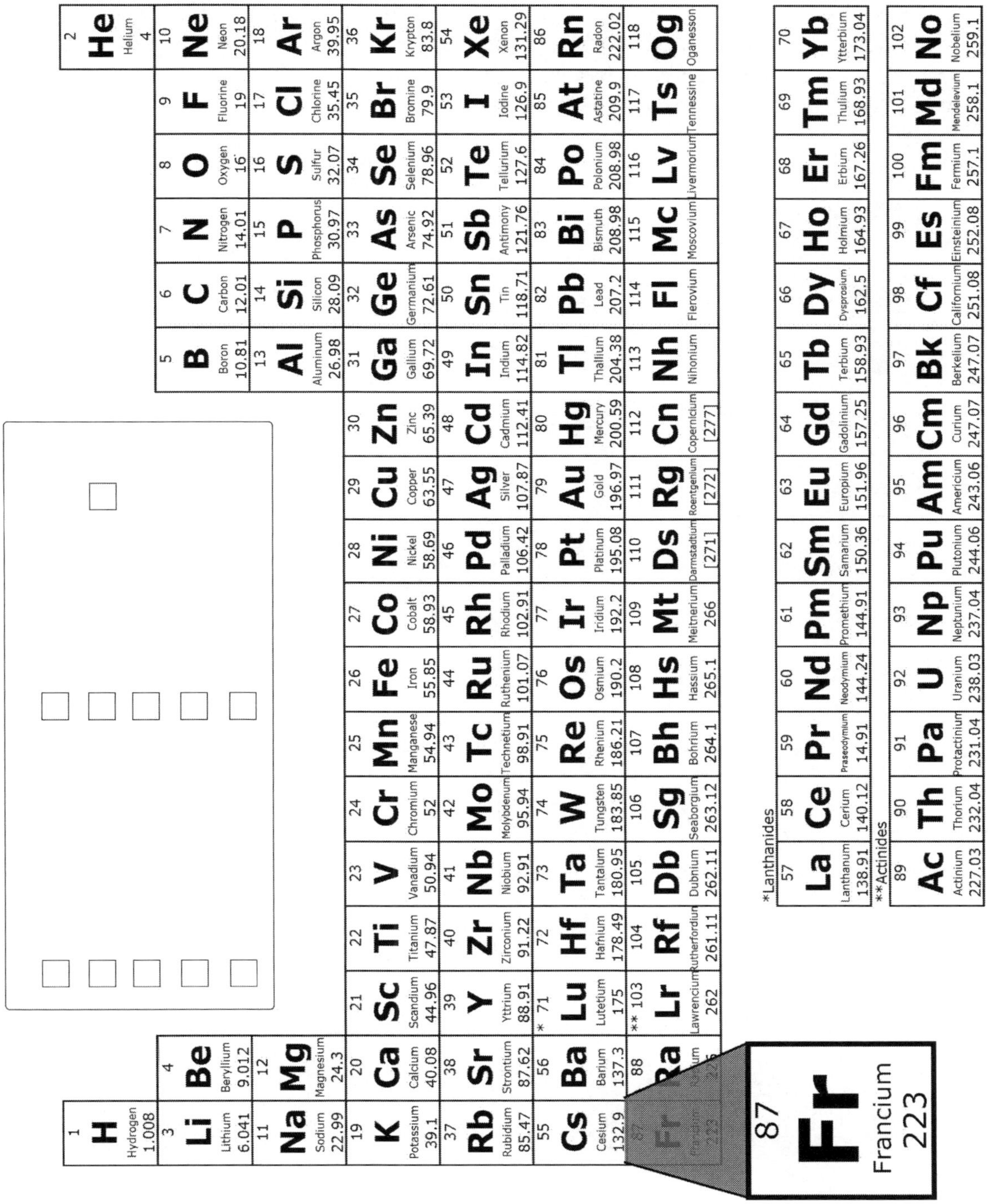

Written Assignment Lesson 2

Discussion Questions

1. How are the elements on the periodic table organized?
2. What was special about Dmitri Mendeleyev's version of the periodic table?
3. How does the periodic table work?
4. What happens when you go across a period?
5. What happens when you go down a group?

Written Assignment Lesson 2

Student Assignment Sheet Lesson 3
Metals

Experiment: Can I transfer metal atoms?

Materials:

- ✓ Vinegar
- ✓ Salt
- ✓ 6 pennies
- ✓ Glass cup
- ✓ 2 iron nails

Procedure:

1. Read the introduction to this experiment and fill in your hypothesis.
2. Cover the bottom of your cup with a thin layer of salt. Add the pennies and cover them with vinegar.
3. After 10 minutes, take the pennies out and set them on a paper towel. Be sure to reserve the vinegar mixture.
4. Add one of the nails to the vinegar solution and leave the other one next to the glass. Let both of them sit undisturbed for 45 minutes. While you are waiting, observe what is happening to the pennies and write down what you see on your experiment sheet.
5. Take the nail out of the vinegar solution and compare it to the nail that was left outside the glass. Write your observations down on your experiment sheet.
6. Draw conclusions and complete your experiment sheet.

Vocabulary & Memory Work

- ☐ Vocabulary: metal
- ☐ Memory Work—This week, add the following elements to what you are working on memorizing:
  - ✓ 9-F-Fluorine, 10-Ne-Neon, 11-Na-Sodium, 12-Mg-Magnesium

Sketch: Metals of the Periodic Table

- Label the Following: the atomic number and symbol for the alkali metals, the alkaline-earth metals and the transition metals

Writing

- Reading Assignment: *DK Encyclopedia of Science* pg. 34 (Alkali Metals), pg. 35 (Alkaline-earth Metals) and pp. 36-37 (Transition Metals)
- Additional Research Readings:
  - Metals: *KSE* pp. 183
  - The Alkali Metals: *UDIS* pp. 168-169
  - The Alkaline-earth Metals: *UDIS* pp. 170-171
  - The Transition Metals: *UDIS* pp. 172-175
  - Looking at Metals: *DK Chem* pp. 24-25

Dates

- 4000 BC – Metals like lead and silver are melted down from ores and used.

Sketch Lesson 3

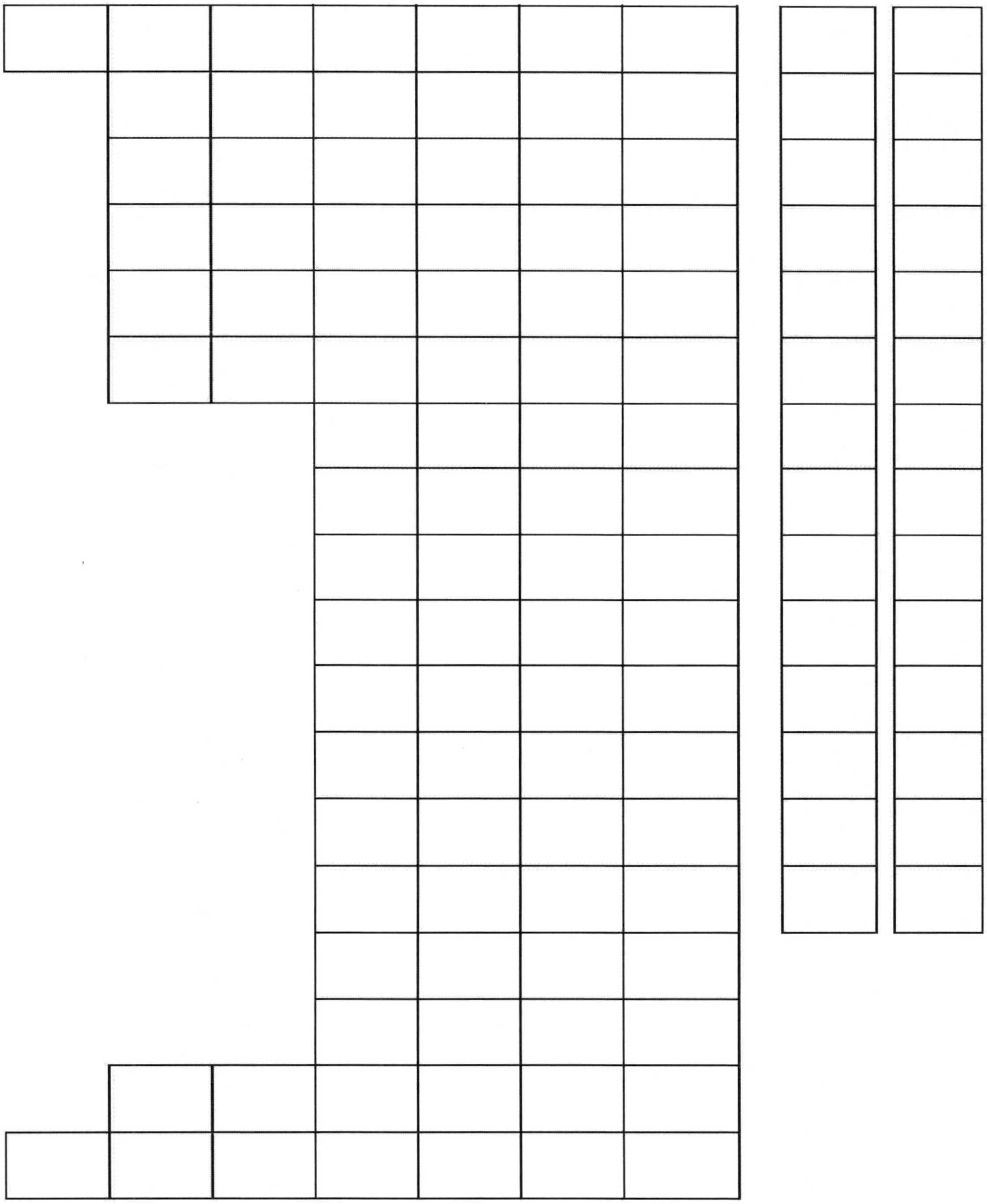

Experiment: Can I transfer metal atoms

Introduction

Copper is a metal that has many uses. It is found in the Earth's crust in a pure state. This metal is an excellent conductor, which is why it is used in wiring. It is also used in roofing, coins and decorative statues. Over prolonged exposure to the air, it will develop a dull green finish, like the Statue of Liberty. In this experiment, you are going to see if you can transfer copper atoms to an iron nail.

Hypothesis

Can I transfer metal atoms? Yes No

Materials

______________________ ______________________

______________________ ______________________

______________________ ______________________

______________________ ______________________

______________________ ______________________

Procedure

____

____

____

____

____

____

____

____

____

____

____

____

Observations and Results

Picture of My Experiment

Conclusion

Written Assignment Lesson 3

Discussion Questions

pg. 34, Alkali Metals

1. What are several of the uses for alkali metals?

pg. 35, Alkaline-earth Metals

1. What are some of the uses of alkaline-earth metals?

pp. 36-37, Transition Metals

1. What are some of the typical characteristics of transition metals?
2. Where are transition metals found?

Written Assignment Lesson 3

Student Assignment Sheet Lesson 4
Poor Metals and Semimetals

Experiment: Is it a metal?

Materials:

- ✓ Magnet
- ✓ Materials for circuit (see Appendix pg. 332)
- ✓ Hammer
- ✓ Charcoal
- ✓ Metal paperclip (not plastic coated)
- ✓ Aluminum foil
- ✓ CD
- ✓ Safety glasses

Procedure:

1. Read the introduction to this experiment and fill in your hypothesis.
2. Conduct the four tests below for each of your materials. Fill in the chart on your experiment sheet as you go.
   - **Luster Test** – Hold your sample material up to the light. Observe whether or not it appears to shine. Does it reflect the light?
   - **Magnetism Test** – Hold a magnet close to your sample material. Is the sample attracted to the magnet?
   - **Conductivity Test** – Build the circuit using the directions found on pg. 332 in the appendix of this guide. Place your sample material in the path of the circuit. Does the light bulb light up?
   - **Malleability Test** – Take your sample material outside and place it on a hard, safe surface. After you put on your safety glasses, strike the material with a hammer and observe what happens. Was your sample flattened by the hammer or did it break apart?

   If you answered yes to all four tests, your material is most likely a metal. If you answered no to all the questions your sample material is most likely a nonmetal. If you answered both yes and no to the questions, your sample material is either a poor metal or a semimetal.
3. Draw conclusions and complete your experiment sheet.

Vocabulary & Memory Work

- ☐ Vocabulary: poor metal, semimetal, malleable, semiconductor
- ☐ Memory Work—This week, add the following elements to what you are working on memorizing:
  - ✓ 13-Al-Aluminum, 14-Si-Silicon, 15-P-Phosphorus, 16-S-Sulfur

Sketch: Poor Metals and Semimetals of the Periodic Table

- Label the following: the atomic number and symbol for the poor metals and semimetals, then color the poor metals blue and the semimetals green

Writing

- Reading Assignment: *DK Encyclopedia of Science* pg. 38 (Poor Metals) and pg. 39 (Semimetals)

Dates

- 1954 – Morris Tanenbaum invents the first silicon transistor at Bell Labs, beginning the use of this semiconductor in electronics.

Sketch Lesson 4

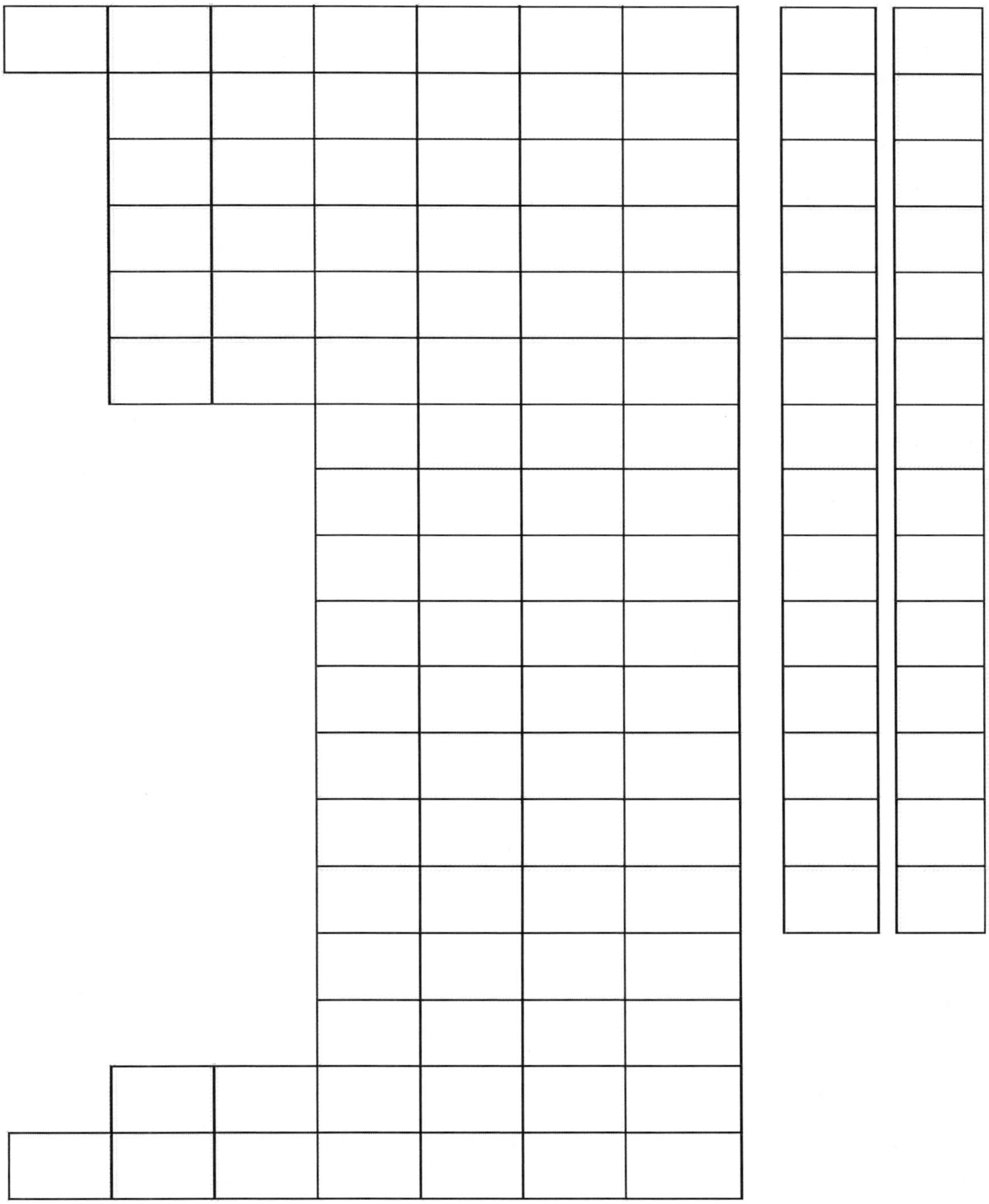

Experiment: Is is a metal?

Introduction

Metals are typically shiny and hard. They are also good conductors of electricity and heat. Some metals can be magnetic and some are malleable. On the other hand, nonmetals are typically brittle, dull and don't conduct electricity well. Poor metals and semimetals have one or more of these characteristics which places them somewhere in between metals and a nonmetals. In this experiment, you are going to use 4 tests to determine if a variety of materials are metals, nonmetals, or somewhere in between.

Hypothesis

Circle the materials below that you believe are metals:

charcoal CD aluminum foil paper clip

Materials

________________ ________________

________________ ________________

________________ ________________

________________ ________________

________________ ________________

Procedure

____

____

____

____

____

____

____

____

____

____

Observations and Results

Material	Luster Test	Magnetism Test	Conductivity Test	Malleability Test
Charcoal				
Paperclip				
Aluminum Foil				
CD				

Conclusion

Written Assignment Lesson 4

Discussion Questions

pg. 38, Poor Metals

1. What are the main characteristics of poor metals?
2. What are some of the uses of poor metals?

pg. 39, Semimetals

1. What are the main characteristics of semimetals?
2. What are some of the ways the semimetals are used?

Written Assignment Lesson 4

Student Assignment Sheet Lesson 5
Nonmetals

Activity: The Periodic Table Game

Materials:

- ✓ Element cards

Procedure:

1. Make an index card for all of the elements of the periodic table, if you have not already done so. Each element card should include the element's name, abbreviation, its atomic number and mass. The card should also say who discovered the element, when they discovered it, and some common uses of the element. You can choose to write out the uses or illustrate them. (***Note:*** *See your instructor for a template to use when making your cards.*)
2. Once you complete the set you can play Guess the Element or Periodic Table Bingo. (***Note:*** *See your instructor for directions for the games.*)

Vocabulary & Memory Work

- ☐ Vocabulary: nonmetal
- ☐ Memory Work—This week, add the following elements to what you are working on memorizing:
  - ✓ 17-Cl-Chlorine, 18-Ar-Argon, 19-K-Potassium, 20-Ca-Calcium

Sketch: Nonmetals of the Periodic Table

- Label the following: the atomic number and symbol for the nonmetals and noble gases, then color the nonmetals green and the noble gases purple

Writing

- Reading Assignment: *DK Encyclopedia of Science* pg. 42 (Nitrogen), pg. 43 (Phosphorus), pg. 44 (Oxygen), pg. 45 (Sulfur) and pg. 46 (Halogens)
- Additional Research Readings:
  - Halogens: *UIDS* pp. 186-188, *KSE* pg. 182
  - Noble Gases: *UIDS* pg. 189, *KSE* pg. 180
  - Discovering Nonmetals: *DK Chem* pp. 26-27

Dates

- 1722 – Antoine Lavoisier shows that diamonds are a form of carbon by burning samples of charcoal and diamond. He fings that neither produced any water and that both released the same amount of carbon dioxide per gram.
- 1772 & 1774 – Swedish chemist Carl Scheele and English chemist Joseph Priestley both discover oxygen.

Sketch Lesson 5

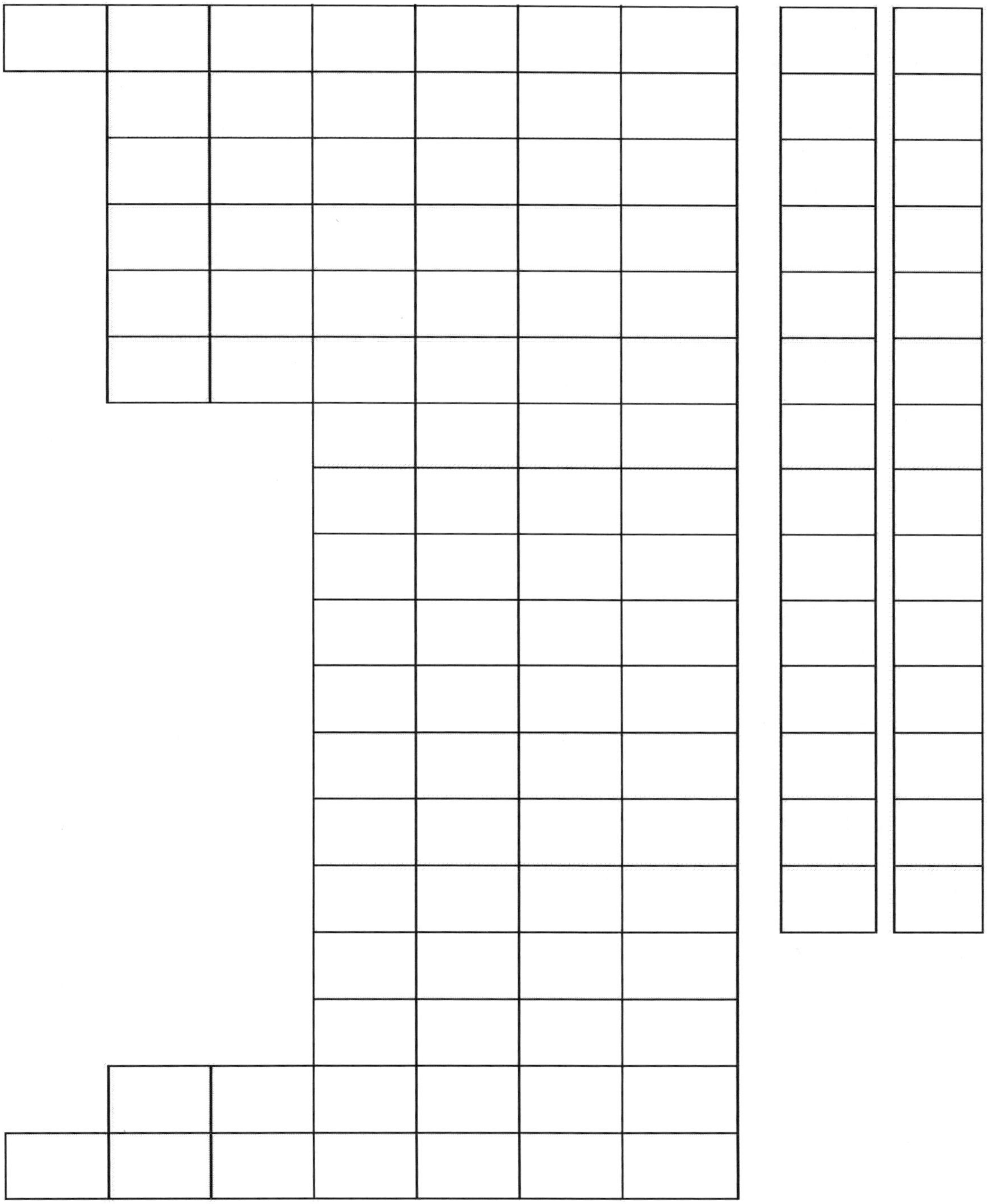

Written Assignment Lesson 5

Discussion Questions

pg. 42, Nitrogen

1. What is the nitrogen cycle?

pg. 43, Phosphorus

1. Describe the common form of phosphorus.

pg. 44, Oxygen

1. Where can oxygen be found?
1. What are three reactions that require the presence of oxygen?

pg. 45, Sulfur

1. Why is sulfur important to industry?

pg. 46, Halogens

1. What are two characteristics common to all halogens?

Written Assignment Lesson 5

Unit 4

Matter

Unit 4: Matter
Vocabulary Sheet

Define the following terms as they are assigned on your Student Assignment Sheet.

1. Plasma – ____

____

____

2. Sublimation – ____

____

____

3. Freezing Point – ____

____

____

4. Boiling Point – ____

____

____

5. Density – ____

____

____

6. Conductivity – ____

____

____

7. Elasticity – ____

____

____

8. Diffusion – ____

____

____

9. Kinetic Theory – ______________________________

10. Pressure – ______________________________

11. Temperature – ______________________________

12. Volume – ______________________________

13. Cleavage – ______________________________

14. Lattices – ______________________________

Student Assignment Sheet Lesson 1
States of Matter

CAUTION
DO NOT DO THIS EXPERIMENT WITHOUT ADULT SUPERVISION!
Do not touch the burner! Be sure to use protective gear when working near the burner.

Experiment: When does water change state?

Materials:
- ✓ Cup
- ✓ Ice cubes
- ✓ Pot
- ✓ Thermometer

Procedure:
1. Read the introduction to this experiment and make a hypothesis.
2. Fill a small pot halfway with ice cubes. Place the pot on a burner and turn the burner onto medium heat. Observe the thermometer as the ice begins to melt and record the temperature once all of the ice melts.
3. Continue to heat the water, observing the temperature on the thermometer as it heats up. Once you begin to see the water boiling and observe the presence of steam, record your last temperature measurement.
4. Turn the burner off and remove the pot from the burner before you draw conclusions and complete your experiment sheet.

Vocabulary & Memory Work
- ☐ Vocabulary: plasma, freezing point, boiling point, sublimation
- ☐ Memory Work—This week, add the following elements to what you are working on memorizing:
  - ✓ 21-Sc-Scandium
  - ✓ 22-Ti-Titanium
  - ✓ 23-V-Vanadium
  - ✓ 24-Cr-Chromium

Sketch: States of Matter
- Label the following: gas, liquid, solid, plasma, freezing, melting, condensing, evaporating, sublimation, deposition, ionization, deionization

Writing
- Reading Assignment: *DK Encyclopedia of Science* pp. 18-19 (States of Matter) and pp. 20-21 (Changes in State)
- Additional Research Readings:
  - States of Matter: *KSE* pp. 156-157
  - States of Matter: *UDIS* pp. 120-121

Dates
- 1897 – Sir William Crookes discovers a fourth state of matter through his experiments with gases.
- 1928 – US scientist Irving Langmuir names the fourth state of matter plasma.

Sketch Lesson 1

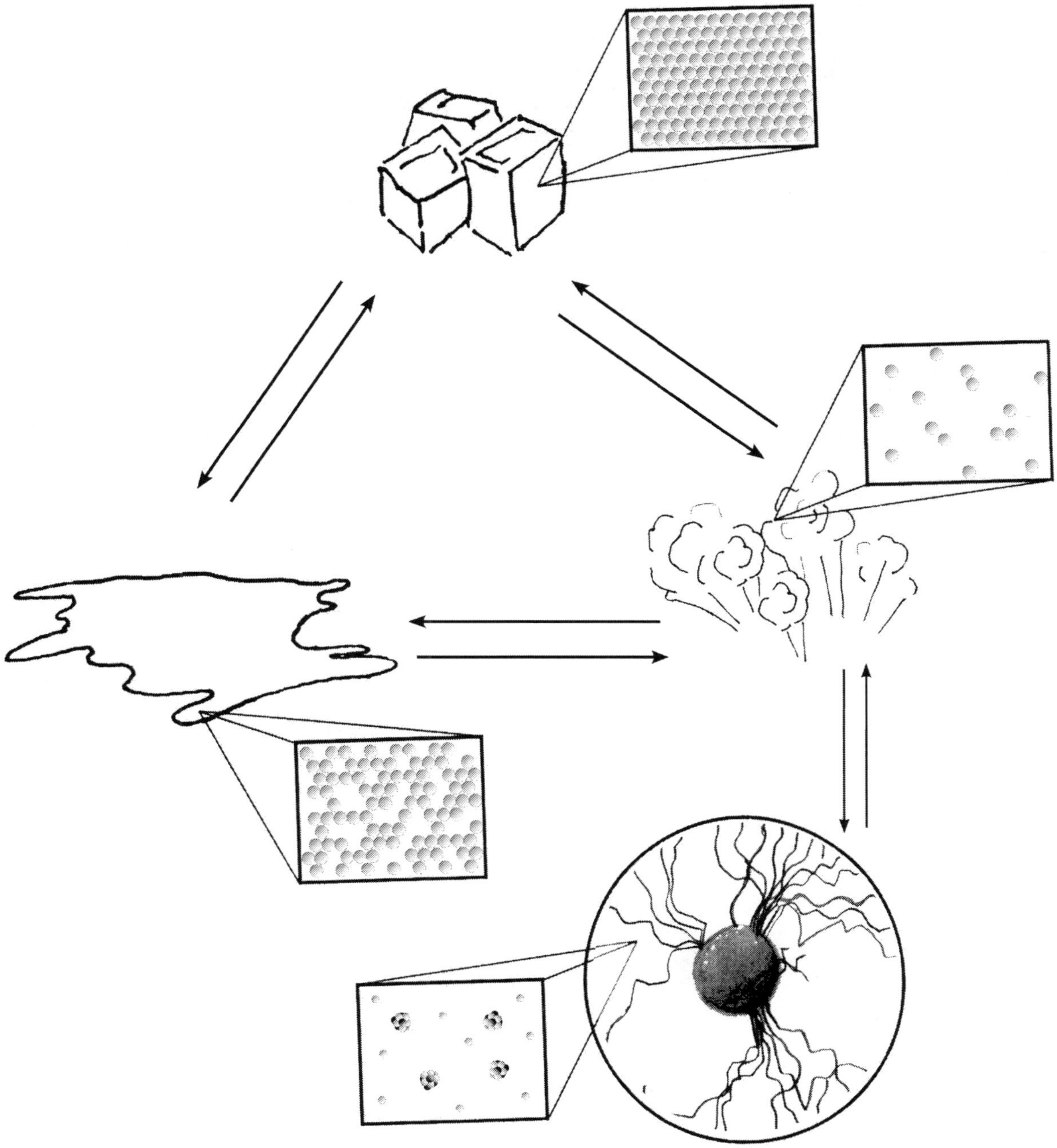

Experiment: When does water change state?

Introduction

There are three states of matter that are easily observed every day: solid, liquid and gas. Solids have tightly packed molecules with fixed shape and volume, liquids have widely spaced molecules with a fixed volume and gases have independently moving molecules with no fixed shape or volume. There is also a fourth state of matter called plasma, which is rarely seen on Earth. Plasma is created when a gas ionizes at very high temperatures. In this experiment, you are going to look at the points at which water changes state.

Hypothesis

I think that I will see the ice change into liquid water at _________.

I think that I will see the liquid water change into gas at _________.

Materials

_______________________________ _______________________________

_______________________________ _______________________________

_______________________________ _______________________________

_______________________________ _______________________________

_______________________________ _______________________________

Procedure

____

____

____

____

____

____

____

____

____

Observations and Results

Phase Change	Temperature
solid to liquid	
liquid to gas	

Conclusion

Written Assignment Lesson 1

Discussion Questions

pp. 18-19, States of Matter

1. What is a solid?
2. What is a liquid?
3. What is a gas?
4. What is plasma?

pp. 20-21, Changes in State

1. What is condensation? Evaporation?
2. What is melting? Freezing?
3. What is sublimation?
4. How does the water cycle relate to states of matter?

Written Assignment Lesson 1

Student Assignment Sheet Lesson 2
Properties of Matter

Experiment: How does the addition of salt affect the density of water?

Materials:

- ✓ 4 clear cups
- ✓ Eye dropper
- ✓ Table salt
- ✓ Food coloring
- ✓ Water

Procedure:

1. Read the introduction to this experiment and make a hypothesis. Then, prepare your solutions according to the directions on your experiment sheet.
2. Label each of your cups #1-4. Fill cup #1 with 1 cup (240 mL) of the yellow solution, cup #2 with 1 cup (240 mL) of the red solution, cup #3 with 1 cup (240 mL) of the blue solution, and cup #4 with 1 cup (240 mL) of the green solution.
3. Using an eye dropper, draw up about 1-2 tsp (5-10 mL) of the yellow solution and ***very slowly*** add it to the cup #2. (***Note****—When you add the yellow solution, you need to make sure the tip of your eyedropper is just below the surface of the solution in the cup.*) Wait 1 minute and record your observations.
4. Next, using an eye dropper, draw up about 1-2 tsp (5-10 mL) of the yellow solution and ***very slowly*** add it to the cup #3. Wait 1 minute and record your observations.
5. Then, using an eye dropper, draw up about 1-2 tsp (5-10 mL) of the yellow solution and ***very slowly*** add it to the cup #4. Wait 1 minute and record your observations.
6. Pour out the solutions from each of the used cups and refill them with 1 cup of each of the colors as before.
7. Repeat steps 3 through 6 with the red, blue and green solution. (***Note****—Each time you will skip the cup with the same color solution because we already know that they have the same concentration of salt.*)
8. Draw conclusions and complete your experiment sheet.

Vocabulary & Memory Work

- ☐ Vocabulary: density, conductivity, elasticity
- ☐ Memory Work—This week, add the following elements to what you are working on memorizing:
  - ✓ 25-Mn-Manganese, 26-Fe-Iron, 27-Co-Cobalt, 28-Ni-Nickel

Sketch: Density Column

- Based on what you know, label the 5 layers of the density column with these liquids: water (density 1.00 g/mL), corn syrup (density 1.38 g/mL), salt water (density 1.03 g/mL), vegetable oil (density 0.92 g/mL), and rubbing alcohol (density 0.79 g/mL)

Writing

- Reading Assignment: *DK Encyclopedia of Science* pp. 22-23 (Properties of Matter)
- Additional Research Readings: Density *UIDS* pg. 24

Dates

- 287 BC-212 BC – Archimedes lives. He uses the principle of density to determine if King Hiero II's crown was made of real gold.

Sketch Lesson 2

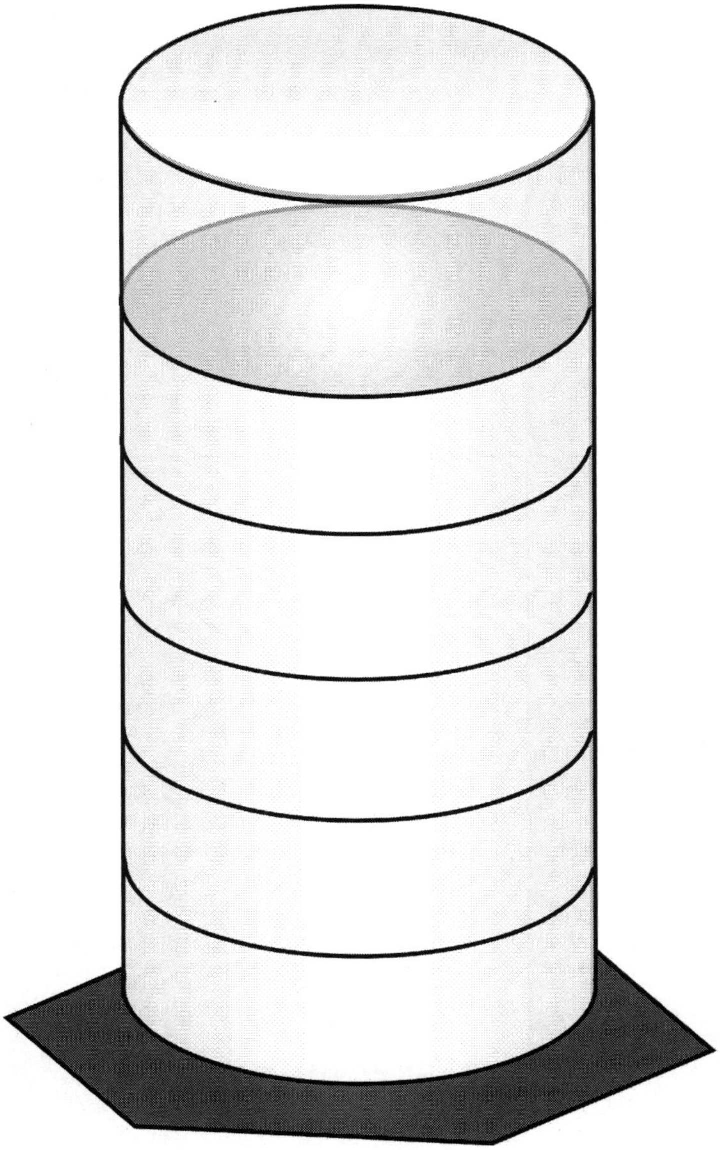

Experiment: How does the addition of salt affect the density of water?

Introduction

Density is a property of matter that relates mass and volume. Mass tells you how much matter is in something, while volume tells you how much space something occupies. So in other words, density measures how much matter of a particular substance will occupy a given space. When comparing two different solutions, a more dense solution will sink to the bottom, while a less dense solution will float to the top. In this week's experiment, you are going to use the principle of density to see how salt affects the density water.

Hypothesis

I think that the addition of salt to the water will (raise / lower) the density.

Materials

______________________ ______________________

______________________ ______________________

______________________ ______________________

How to prepare your solutions:

1. Begin by putting 4 cups (1000 mL) of room temperature water into 4 separate bowls.
2. For the yellow solution, add 10 drops of yellow food coloring to one of the bowls.
3. For the red solution, add 10 drops of red food coloring and 2 TBSP (34.2 g) of salt to another bowl.
4. For the blue solution, 10 drops of blue food coloring and ¼ cup (68.3 g) of salt to another bowl.
5. For the green solution, add 10 drops of green food coloring and ½ cup (136.5 g) of salt to last bowl.
6. Stir each one to make sure that the salt is thoroughly dissolved.

Procedure

____

____

____

____

____

Observations and Results

	Added to the Yellow Solution	Added to the Red Solution	Added to the Blue Solution	Added to the Green Solution
Yellow Solution				
Red Solution				
Blue Solution				
Green Solution				

Conclusion

Written Assignment Lesson 2

Discussion Questions

1. Why do scientists measure the properties of different types of matter at the same temperature and pressure?
2. What is the difference between mass and volume?
3. What makes a good conductor? Insulator?
4. Why is water called the universal solvent?

Written Assignment Lesson 2

Student Assignment Sheet Lesson 3
Kinetic Theory and Gases

> ☢ **CAUTION**
> **DO NOT DO THIS EXPERIMENT WITHOUT ADULT SUPERVISION!**
> The apple juice could be very hot when removing it from the microwave. Make sure to use protective gear.

Experiment: Does temperature affect the rate of diffusion?

Materials:

- ✓ 2 cups, apple juice, timer, partner

Procedure:

1. Read the introduction to this experiment and make a hypothesis.
2. Fill 2 cups with apple juice, cover each of them with plastic wrap, and leave them on the counter.
3. After one hour, take one of the cups from the counter and have your partner stand in the opposite corner of the room. Remove the plastic wrap from the cup, time how long it takes for them to smell the apple juice. Record that time on your experiment sheet. (***Note****—If your partner doesn't smell it after 10 minutes, end the test and note the results on the experiment sheet.*)
4. Next, take the other cup from the counter and microwave it on high for 1-2 minutes, until the juice is warm, but not too hot to touch. Have your partner stand in the opposite corner of the room and remove the plastic wrap from the cup. Time how long it takes for them to smell the apple juice and record that time on your experiment sheet. (***Note****—If your partner doesn't smell it after 10 minutes, end the test and note the results on the experiment sheet.*)
5. Draw conclusions and complete your experiment sheet.

Vocabulary & Memory Work

- ☐ Vocabulary: diffusion, kinetic theory, pressure, temperature, volume
- ☐ Memory Work—This week, add the following elements to what you are working on memorizing:
  - ✓ 29-Cu-Copper, 30-Zn-Zinc, 31-Ga-Gallium, 32-Ge-Germanium
- ☐ Memory Work—Work on memorizing Boyle's Law and Charles' Law (*see the pg. 53 for these gas laws*).

Sketch: Gas Laws

- Label the Following – Boyle's Law: volume decrease, pressure increase, constant temperature; Charles' Law: temperature increase, volume increase, constant pressure.

Writing

- Reading Assignment: *DK Encyclopedia of Science* pg. 50 (Kinetic Theory) and pg. 51 (Behavior of Gases)
- Additional Research Readings:
  - 🕮 Looking at Air: *DK Chem* pg. 28-29, Gas Laws: *UDIS* pg. 142-143, Air: *KSE* pg. 172

Dates

- 🕒 1827 – Robert Brown observes pollen grains randomly bouncing round in water, discovering Brownian Motions, which happen because the grains are being constantly hit by water molecules.
- 🕒 1860 – Ludwig Boltzmann develops the kinetic theory of gases, which is widely opposed by other scientists.

Sketch Lesson 3

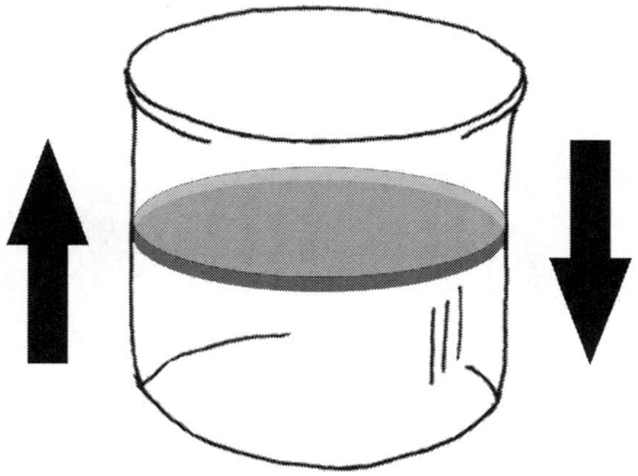

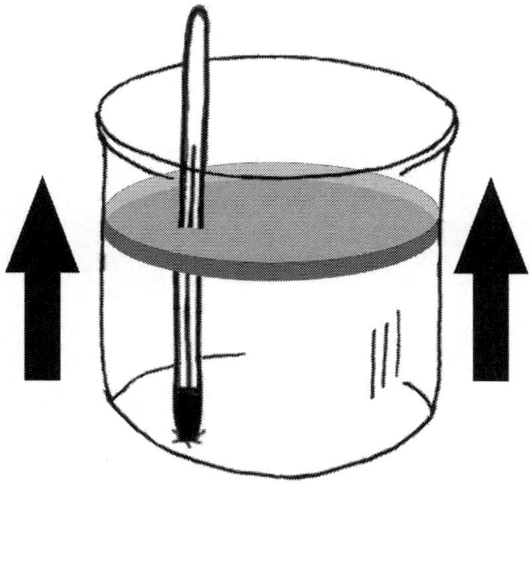

Experiment: Does temperature affect the rate of diffusion?

Introduction

The particles in a substance are constantly in motion. In a solid, these particles are closely packed together, so they can only move by vibrating or shaking. In a liquid, these particles can move more freely, but they still remain relatively close. In a gas, these particles are widely spaced and can move very quickly. When the particles in a substance move from one corner of a room to the other, they are said to be diffusing, or spreading out. In this week's experiment, you are going to examine how temperature affects the rate of diffusion.

Hypothesis

I think that temperature (will / will not) affect the rate of diffusion.

Materials

______________________________ ______________________________

______________________________ ______________________________

______________________________ ______________________________

______________________________ ______________________________

______________________________ ______________________________

Procedure

____

____

____

____

____

____

____

____

____

____

Observations and Results

Sample	Time it took for the smell to reach my partner
Room Temperature Apple Juice	
Hot Apple Juice	

Conclusion

Written Assignment Lesson 3

Discussion Questions

pg. 50, Kinetic Theory

1. What is common to all atoms and molecules?
2. What causes diffusion in gases?
3. Why do two liquids diffuse (or mix together)?
4. What happens during expansion?

pg. 51, Behavior of Gases

1. Why can changing the temperature, volume or pressure of a gas have dramatic results on the behavior of the gas?
2. What is an example of Boyle's Law?
3. What is an example of Charles' Law?
4. What is Avogadro's Law?

Written Assignment Lesson 3

Student Assignment Sheet Lesson 4
Solid Structures

Experiment: Crystals

> ☢ **CAUTION**
> **DO NOT DO THIS EXPERIMENT WITHOUT ADULT SUPERVISION!**
> Boiling hot water can cause severe damage. Be sure to use the proper safety gear.

Materials:
- ✓ Wide mouthed jar
- ✓ String, Pencil, Pipe cleaner
- ✓ Water, Borax
- ✓ Scissors

Procedure:
1. Read the introduction to this experiment.
2. Fill your jar with water, pour that water into a pot and set it on a burner to boil. Make sure to record how many cups (or mL) of water it takes to fill your jar.
3. Form a shape with the pipe cleaner. This can be as simple or as complex as you wish, but make sure it will fit through the opening of your jar. Next, tie a string to the shape and then tie the other end of the string to a pencil. (***Tip****—You want the pencil to be able to rest on the edge of your jar without having your shape touch the sides or bottom of the jar.*)
4. Set your jar in the sink, put on saftey glasses and use a hot mitt to slowly add boiling water until it almost fills the jar. Then, add the Borax one TBSP (25 g) at a time, stirring each time until the Borax is dissolved. You want to add about 3 TBSP (75 g) of Borax for every cup (240 mL) of water you've added. (***Note****—Be sure to record how many tablespoons you added on your experiment sheet.*)
5. Hang the shape in the jar so that it is completely covered by the liquid. Allow the jar to sit undisturbed overnight. In the morning, take out your shape and observe what has happened.
6. Draw conclusions and complete your experiment sheet.

Vocabulary & Memory Work
- ☐ Vocabulary: cleavage, lattices
- ☐ Memory Work—This week, add the following elements to what you are working on memorizing:
  - ✓ 33-As-Arsenic, 34-Se-Selenium, 35-Br-Bromine, 36-Kr-Krypton

Sketch: Crystal Systems
- Label the following: cubic symmetry, rhombic symmetry, hexagonal symmetry, tetragonal symmetry, monoclinic symmetry, triclinic symmetry and trigonal symmetry

Writing
- Reading Assignment: *DK Encyclopedia of Science* pg. 30 (Crystals)
- Additional Research Readings:
  - Crystals: *UIDS* pp. 135-137
  - Crystals: *KSE* pp. 168-169

Dates
- 1915 – William Bragg and his son receive the Nobel Prize in Physics for their work on crystalline structures.

Sketch Lesson 4

Experiment: Can I make crystals?

Introduction

Crystals come in all shapes and sizes, but each is based on the regularly repeating pattern of the atoms, molecules or ions that make it up. They are generally very hard solid structures and we use them in a variety of ways as we go about our daily life. For example, the salt you use on your food and the lead in your pencil are both made up of crystals. In this experiment, you are going to attempt to make your own crystals.

Materials

Procedure

Observations and Results

Picture of My Experiment

Conclusion

Written Assignment Lesson 4

Discussion Questions

1. What causes crystals to form a definite shape?
2. What gives certain crystals, such as silicone dioxide, a range of colors?
3. What do scientists measure to help them identify crystals?
4. How does a liquid crystal display work?

Written Assignment Lesson 4

Unit 5

Solutions

Unit 5: Solutions
Vocabulary Sheet

Define the following terms as they are assigned on your Student Assignment Sheet.

1. Molecule – ______________________________

____

____

2. Mixture – ______________________________

____

____

3. Miscible – ______________________________

____

____

4. Immiscible – ______________________________

____

____

5. Solution – ______________________________

____

____

6. Solubility – ______________________________

____

____

7. Solute – ______________________________

____

____

8. Solvent – ______________________________

____

____

9. Anode – ____

____

____

10. Cathode – ____

____

____

11. Electrolyte – ____

____

____

12. Chromatography – ____

____

____

Student Assignment Sheet Lesson 1
Molecules and Mixtures

Experiment: Marshmallow Molecules

Materials:

- ✓ Bag of multi-colored marshmallows
- ✓ Toothpicks

Procedure:

1. Read the introduction to this experiment.
2. Choose a color of marshmallow to represent each of the following atoms: oxygen, nitrogen, hydrogen and carbon. When you make your molecules you must follow these rules:
   - ➔ Oxygen prefers to bond twice, nitrogen prefers to bond three times, hydrogen prefers to only bond once and carbon prefers to bond 4 times;
   - ➔ Each atom must have its preferred number of bonds to form a stable molecule and you must only create stable molecules;
   - ➔ All of nitrogen's bonds should point down and all of carbon's bonds need to be opposite from each other (except in the case of a multiple bond).
3. Make the following molecules from your marshmallows: NH_3 (ammonia), H_2O (water), CH_4 (methane), CO_2 (carbon dioxide), C_2H_5OH(ethanol) (***Hint****—You can make multiple bonds between the atoms*.)
4. Draw a quick sketch of each of your molecules after you finish assembling them.
5. Draw conclusions and complete your experiment sheet.

Vocabulary & Memory Work

- ☐ Vocabulary: molecule, mixture
- ☐ Memory Work—This week, add the following elements to what you are working on memorizing:
  - ✓ 37-Rb-Rubidium, 38-Sr-Strontium, 39-Y-Yttrium, 40-Zr-Zirconium
- ☐ Memory Work—Work on memorizing the Law of Constant Composition: A pure compound always contains the same elements in the same proportions.

Sketch

- No sketch this week.

Writing

- Reading Assignment: *DK Encyclopedia of Science* pp. 58-59 (Compounds and Molecules)
- Additional Research Readings: Chemical Compounds: *KSE* pp. 164-165, Molecules: *UIDS* pg. 125, Atoms and Molecules: *DK Chem* pp. 16-17

Dates

- 1649 – Pierre Gassendi states that atoms can be joined together to form molecules.
- 1798-1808 – Joseph-Louis Proust analyzes the different sources of several compounds and finds that their elements always contained the same ratio by weight. This leads to the discovery of the law of constant composition.

Experiment: Marshmallow Molecules

Introduction

Molecules are formed when two or more atoms are held together by a chemical bond. These molecules form the millions of substances that surround us in daily life, such as water and table salt. The atoms tend to follow a set of rules when they form molecules. In this experiment, you are going to make marshmallow representations of several common molecules.

Materials

______________________ ______________________

______________________ ______________________

______________________ ______________________

______________________ ______________________

______________________ ______________________

Procedure

____

____

____

Observations and Results

NH_3 (ammonia)

H_2O (water)

CH_4 (methane)

CO_2 (carbon dioxide)

C_2H_5OH(ethanol)

Conclusion

Written Assignment Lesson 1

Discussion Questions

1. How are compounds and mixtures alike? How do they differ?
2. What is the law of constant composition and who discovered it?
3. What are three types of liquid mixtures?
4. What is synthesis?
5. What is an alloy?
6. How do chemists define the purity of a substance?

Written Assignment Lesson 1

Student Assignment Sheet Lesson 2
Solutions

Experiment: Is it polar or nonpolar?

Materials:

- ✓ 5 clear cups (or beakers)
- ✓ 5 plastic spoons
- ✓ Sugar
- ✓ Salt
- ✓ Baking soda
- ✓ Flour
- ✓ Petroleum jelly
- ✓ Water
- ✓ Vegetable oil
- ✓ Tablespoon

Procedure:

1. Read the introduction to this experiment and write down which molecules you think are polar and which ones you think are nonpolar.
2. Label the cups #1 through #5. Add ½ cup (120 mL) of room temperature water to each of the five cups.
3. Then stir in 1 TBSP (12.6 g) of sugar to cup #1, 1 TBSP (17.1 g) of salt to cup #2, 1 TBSP (12 g) of baking soda to cup #3, 1 TBSP (5 g) of flour to cup #4 and 1 TBSP (8 g) of petroleum jelly to cup #5.
4. Gently stir each of the cups with a spoon, using a different spoon for each cup. Wait 10 minutes and record your observations and results.
5. Pour out the solutions, rinse well and dry each of the cups. Add ½ cup (120 mL) of room temperature oil to each of the five cups.
6. Repeat steps 3 and 4.
7. Draw conclusions and complete your experiment sheet.

Vocabulary & Memory Work

- ☐ Vocabulary: solution, solubility, solute, solvent
- ☐ Memory Work—This week, add the following elements to what you are working on memorizing:
  - ✓ 41-Nb-Niobium, 42-Mo-Molybdenum, 43-Tc-Technetium, 44-Ru-Ruthenium
- ☐ Memory Work—Continue to work on the Law of Constant Composition

Sketch: Anatomy of a Water Molecule

- Label the following: chemical bonds, the symbol for an oxygen atom (O), the symbol for the hydrogen atoms (H), the charges associated with the molecule

Writing

- Reading Assignment: *DK Encyclopedia of Science* pg. 60 (Solutions) and pg. 75 (Chemistry of Water)
- Additional Research Readings:
  - Solutions: *KSE* pp. 158-159
  - Solutions and Solubility: *UIDS* pp. 144-145

Dates

- No dates to be entered this week.

Sketch Lesson 2

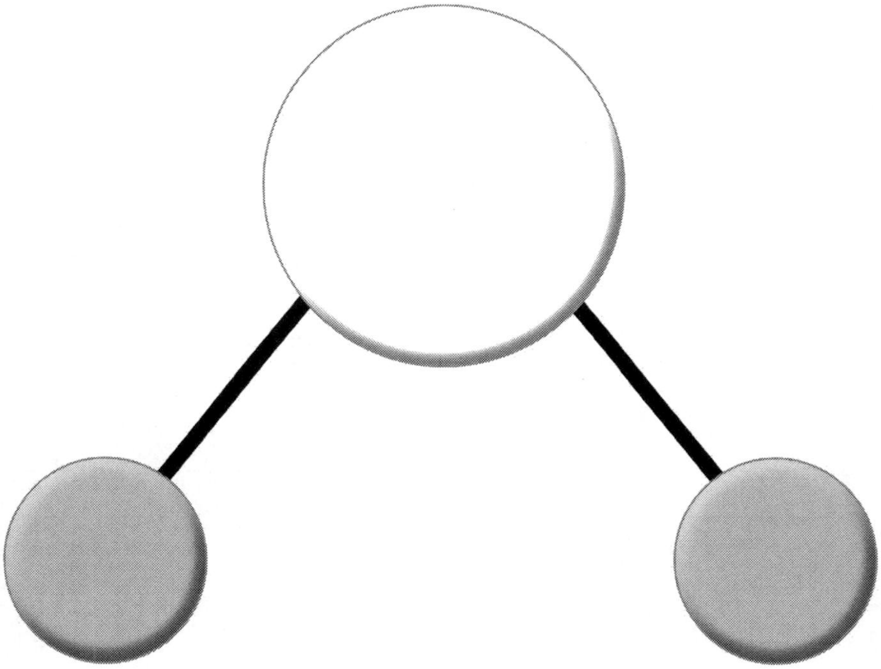

Experiment: Is it polar or nonpolar?

Introduction

A solution contains several different types of molecules that have been evenly mixed. Generally, a solution is made when one type of molecule, which we call the solute, is dissolved in another, which we call the solvent. The molecules can be either polar or nonpolar. Polar molecules have a positive and a negative end, while nonpolar molecules do not carry a charge. As a general rule, like dissolves like; this means that solvents with polar molecules dissolve solutes with polar molecules and vice versa. In this experiment, you are going to use a polar solvent, water, and a nonpolar solvent, vegetable oil, to determine if several substances are polar or nonpolar.

Hypothesis

I think that ______________________________ are polar.

I think that ______________________________ are nonpolar.

Materials

____________________ ____________________

____________________ ____________________

____________________ ____________________

____________________ ____________________

____________________ ____________________

Procedure

____

____

____

____

____

____

____

____

____

Observations and Results

Solute	Solvent: Water	Solvent: Vegetable Oil
Sugar		
Salt		
Baking Soda		
Flour		
Petroleum Jelly		

***S=Soluble, SS=Semi-soluble, NS=not soluble**

____

____

____

____

____

Conclusion

____

____

____

____

____

____

____

Written Assignment Lesson 2

Discussion Questions

pg. 60, Solutions

1. What is the difference between concentrated and dilute solutions?
2. Why is water such a good solvent?
3. What is a saturated solution?

pg. 75, Chemistry of Water

1. Why is water rarely found in its pure state?
2. What is unique to solid water (a.k.a. ice)?
3. What is hard water and what causes it to form?

Written Assignment Lesson 2

Student Assignment Sheet Lesson 3
Separating Mixtures and Compounds

Experiment: Can I split water?

> ☢ **CAUTION**
> **DO NOT DO THIS EXPERIMENT WITHOUT ADULT SUPERVISION!**
> This experiment has the potential to form hydrogen gas, which is extremely flammable.

Materials:
- ✓ Distilled water
- ✓ 2 test tubes
- ✓ Salt
- ✓ Glass cup
- ✓ 2 Alligator clips
- ✓ Covered copper wire
- ✓ 6v Lantern battery
- ✓ Permanent marker

Procedure:
1. Read the introduction to this experiment and make your hypothesis.
2. Mix together 4 cups (1000 mL) of distilled water with ½ cup (136.5 g) of salt and set the solution aside. Next, cut your wire into two lengths and expose the two ends of each of the wires. Attach one end of the first length of wire to one of the alligator clips. Repeat for the second wire.
3. Now, tape your test tubes to opposite sides of your cup. Fill the cup half of the way full with your salt water solution. Some of the water will go into your test tubes, which is what you want. They should be fairly close to full of water. Fit the other exposed end of your wire into the bottom of your test tubes, making sure that it does not touch the sides of your test tube. Rest the remaining part of the wire on the side or your container
4. Then, mark the water level in each of the tubes with a permanent marker and check your set up to make any necessary adjustments. Now, you can attach the alligator clips to each of the battery terminals of the lantern battery. ***CAUTION—Do NOT touch the water when the wire is attached to the battery!!!***
5. Wait for 30 minutes and then detach the wires from the battery. Mark the water level on each of the tubes once again. Remove the tubes and measure the distance between each of your lines to determine how much gas you collected.
6. Draw conclusions and complete your experiment sheet.

Vocabulary & Memory Work
- ☐ Vocabulary: anode, cathode, electrolyte
- ☐ Memory Work—This week, add the following elements to what you are working on memorizing:
  - ✓ 45-Rh-Rhodium, 46-Pd-Palladium, 47-Ag-Silver, 48-Cd-Cadmium
- ☐ Memory Work: Continue to work on the Law of Constant Composition

Sketch: Electrolysis
- Label the Following: anode, cathode, electrolyte solution, battery

Writing
- Reading Assignment: *DK Encyclopedia of Science* pg. 61, Separating Mixtures and pg. 67, Electrolysis
- Additional Research Readings: Separation and Purification: *KSE* pp. 160-161

Dates
- 1778-1829 – English chemist Humphrey Davy lives. He is one of the first scientists to see electrolysis in action.
- 1832 – Michael Faraday writes a mathematical equation that can be used to calculate the quantity of the separated elements in electrolysis.

Sketch Lesson 3

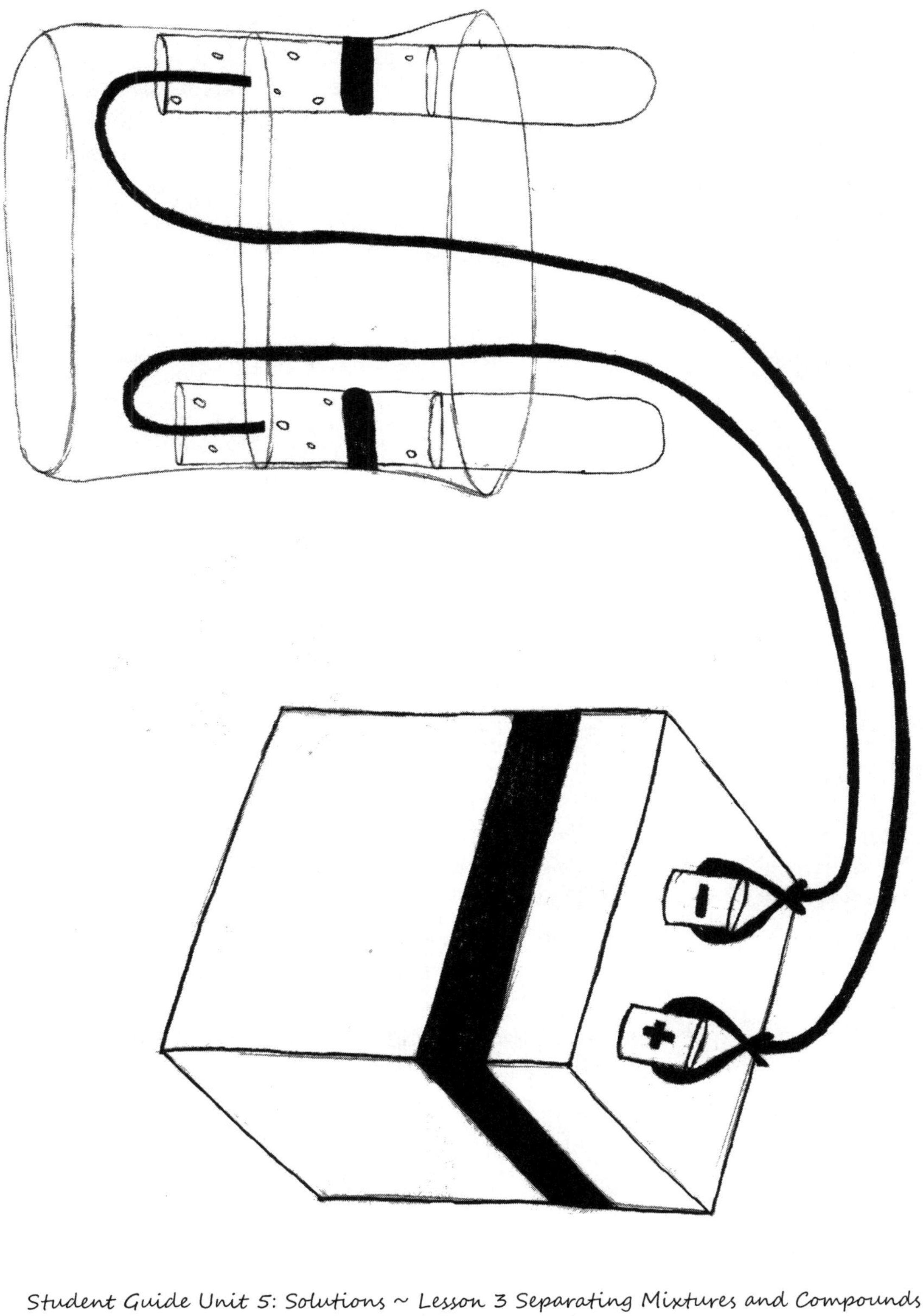

Experiment: Can I split water?

Introduction

Scientists can use a variety of methods to separate mixtures, such as filtration, decantation, distillation, evaporation and centrifugation. Electrolysis separates the components of a molecule within a solution. It uses electrical current to break the compound into its pieces. This process requires that the substance can conduct electricity and that it is either in solution or in a molten state. Electricity passes from two electrodes into the solution, which breaks the compound into its parts. In this experiment we are going to see if we can use electrolysis to split water.

Hypothesis

I think that I will see ____

____

____

Materials

____________________	____________________
____________________	____________________
____________________	____________________
____________________	____________________
____________________	____________________

Caution – Do NOT do this experiment on your own as it has two main hazards, electricity with water and potentially flammable hydrogen gas. It can be done safely under adult supervision in your home as long as you follow the two rules below.

1. **Do NOT touch the water when the wire is connected to the battery.** You can wear rubber gloves when you are dealing with the electrical part of this experiment for an extra measure of safety.
2. **Do NOT do this experiment anywhere near an open flame.** Also, do not smoke in the house while performing this experiment.

Procedure

____

____

Observations and Results

Gas	Amount Collected
Oxygen Gas (found at the anode)	
Hydrogen Gas (found at the cathode)	

Conclusion

Written Assignment Lesson 3

Discussion Questions

pg. 61, Separating Mixtures

1. How does decanting separate mixtures?
2. What is desiccation?
3. What is necessary for filtration to be able to separate the substances in a mixture?
4. What is evaporation?
5. When is distillation generally used to separate a mixture?
6. How does centrifuging separate a mixture?

pg. 67, Electrolysis

1. What creates the negative and positive ions in electrolysis and where do they go?
2. What is electrorefining?
3. How does electroplating use electrolysis?

Written Assignment Lesson 3

Student Assignment Sheet Lesson 4
Chemical Analysis

Experiment: What colors are in a marker?

Materials:

- ✓ 4 coffee filters
- ✓ Markers
- ✓ Alcohol
- ✓ Coffee can or a wide-mouthed jar
- ✓ Rubber bands
- ✓ Eye dropper

Procedure:

1. Read the introduction to this experiment and make your hypothesis.
2. Choose four different colors of marker to test. In the center of one of the coffee filters, color the outline of a circle with a ½ inch (1.2 cm) diameter with one of the markers you chose. Repeat for the remaining markers.
3. Place one of the colored coffee filters over the top of the coffee can with the circle outline in the center of the opening. Use a rubber band to attach the coffee filter to the can.
4. Then, use the eye dropper to drop 7 to 10 drops of rubbing alcohol in the center of the circle outline on the coffee filter. Wait, observe what happens, and record the colors that are left on the filter paper when the alcohol reaches the edge of the can. (**Note**—If the alcohol does not reach the edge of the can, you may need to add a drop or two more.
5. Repeat steps 3 and 4 for the remaining colored coffee filters.
6. Draw conclusions and complete your experiment sheet.

Vocabulary & Memory Work

- ☐ Vocabulary: chromatography
- ☐ Memory Work—This week, add the following elements to what you are working on memorizing:
  - ✓ 49-In-Indium, 50-Sn-Tin, 51-Sb-Antimony, 52-Te-Tellurium

Sketch: Chromatography

- Label the following: sample, solvent, solvent front, filter paper

Writing

- Reading Assignment: *DK Encyclopedia of Science* pp. 62-63 (Chemical Analysis)
- Additional Research Readings:
  - Investigating Substances: *UIDS* pp. 220-221
  - Chemical Analysis: *KSE* pp. 154-155
  - Monitoring Materials: *DK Chem* pp. 60-61

Dates

- 1919 – English chemist Francis Ashton invents the mass spectrometer.

Sketch Lesson 4

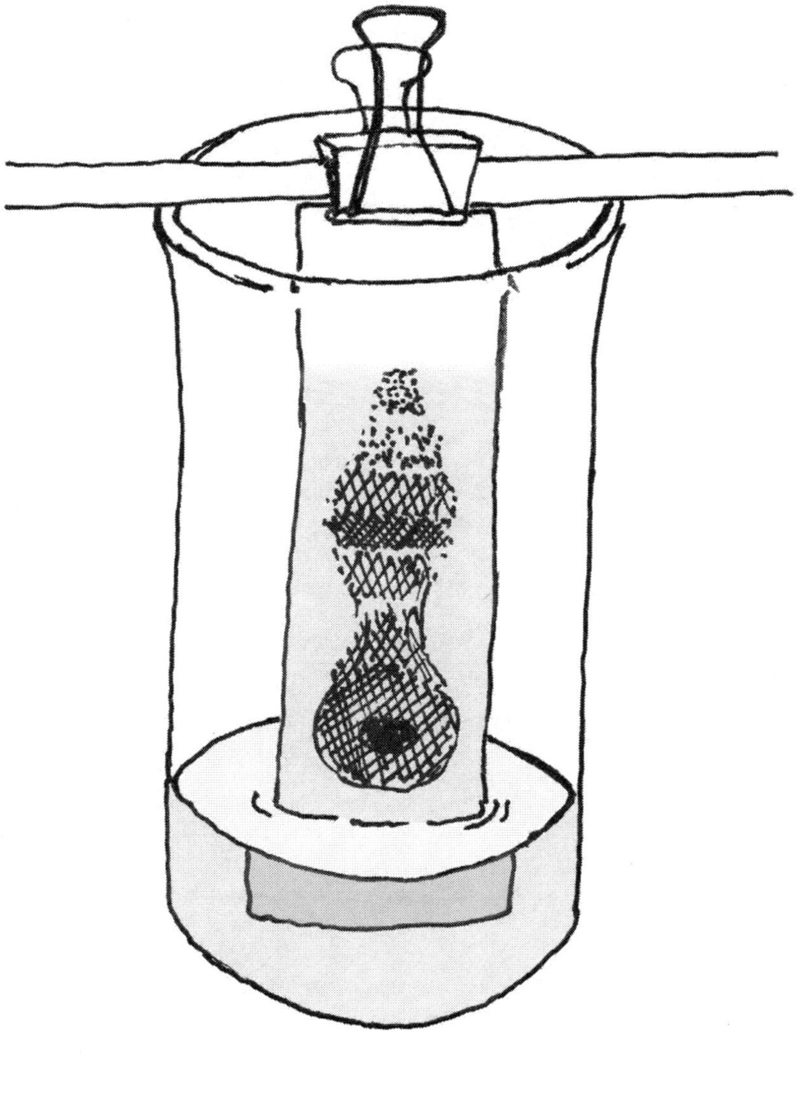

Experiment: What colors are in a marker?

Introduction

Scientists can use several different methods to separate a mixture into its components. One of those ways is called chromatography, which uses a solvent to move the molecules of a mixture through a medium, such as filter paper. The smaller the molecule in a mixture, the farther a solvent will carry it. So, chromatography is able to separate the molecules in a mixture according to their size. In this experiment, you are going to use chromatography to separate the different colors found in a marker.

Hypothesis

I think that I will see ____

____

____

____

Materials

____________________	____________________
____________________	____________________
____________________	____________________
____________________	____________________
____________________	____________________

Procedure

____

____

____

____

____

____

____

____

Observations and Results

Marker Color	Colors Seen

Conclusion

Written Assignment Lesson 4

Discussion Questions

1. What is qualitative analysis?
2. What are two ways that scientists use chemical analysis in real life?
3. Why do scientists use titration?
4. Why do scientists use chromatography?
5. How does a mass spectrometer work?

Written Assignment Lesson 4

Unit 6

Chemical Reactions

Unit 6: Chemical Reactions
Vocabulary Sheet

Define the following terms as they are assigned on your Student Assignment Sheet.

1. Covalent Bonding – ______________________________

2. Ion – ______________________________

3. Ionic Bonding – ______________________________

4. Metallic Bonding – ______________________________

5. Chemical Reaction – ______________________________

6. Endothermic Reaction – ______________________________

7. Exothermic Reaction – ______________________________

8. Product – ______________________________

9. Reactant – ______________________________

10. Moles – ____________________

11. Reactivity – ____________________

12. Concentration – ____________________

13. Equilibrium – ____________________

14. Irreversible Reaction – ____________________

15. Reversible Reaction – ____________________

16. Catalyst – ____________________

17. Enzyme – ____________________

18. Oxidation – ____________________

19. Reduction – ____

____

____

20. Redox Reaction – ____

____

____

Student Assignment Sheet Lesson 1
Chemical Bonding

Experiment: Chocolate Bonding

Materials:

✓ Cake frosting

✓ Red and yellow bite-sized candies (such as regular sized M&M's)

Procedure:

1. Read the introduction to this experiment.
2. **Ionic bonding –** Count out nine red candies and one yellow candy. Arrange the nine red candies in a circle, so that there are two in the center with seven surrounding them. (*Be sure to leave a space for one more in the outer circle.*) These are the electrons in the outer shell of a chlorine atom. Take the one yellow candy, which represents the electron in the outer shell of a sodium atom, and move into the space in the outer circle of the chlorine electrons. Now you have an ionic bond between your sodium and chlorine atom. Draw a picture of your model or take a picture of it for your experiment sheet.
3. **Covalent bonding –** Count out eight red candies and two yellow candies. Use the eight red candies to make a model showing the electrons in the outer shell of an oxygen atom by making a circle with six candies on the outside and two on the inside. (*Be sure to leave space for two more candies in the outer layer.*) Now take two of the yellow candies, which represent the electrons in the outer shells of the two different hydrogen atoms, and bite or cut them in half. Do the same with the two of the red candies from the outer circle you created. Then, use the frosting to glue together one half of each color to form four whole candies. Place the dual colored candies in the outer circle of your model to represent the covalent bond that is found between oxygen and hydrogen in water. Draw a picture of your model or take a picture of it and then complete your experiment sheet.

Vocabulary & Memory Work

- ☐ Vocabulary: covalent bonding, ion, ionic bonding, metallic bonding
- ☐ Memory Work—This week, add the following elements to what you are working on memorizing:
  - ✓ 53-I-Iodine, 54-Xe-Xenon, 55-Cs-Cesium, 56-Ba-Barium

Sketch: Types of Chemical Bonding

- Label the following on the Ionic Bond: donated electron, sodium ion, chlorine ion
- Label the following on the Covalent Bond: shared electrons, oxygen atom, hydrogen atom
- Label the following on the Metallic Bond: fixed positive gold ions, free electrons

Writing

- Reading Assignment: *DK Encyclopedia of Science* pp. 28-29 (Bonding)
- Additional Research Readings: Bonding and Valency: *KSE* pp. 166-167, Atomic Structure: *UIDS* pp. 126-127, Investigating Compounds: *DK Chem* pp. 20-21

Dates

- 1704 – Isaac Newton roughly outlines atomic theory.
- 1916 – Gilbert Lewis develops the concept of electron pair bonding.
- 1954 – Linus Pauling is awarded the Nobel Prize for Chemistry for his work on calculating the energy to break bonds.

Sketch Lesson 1

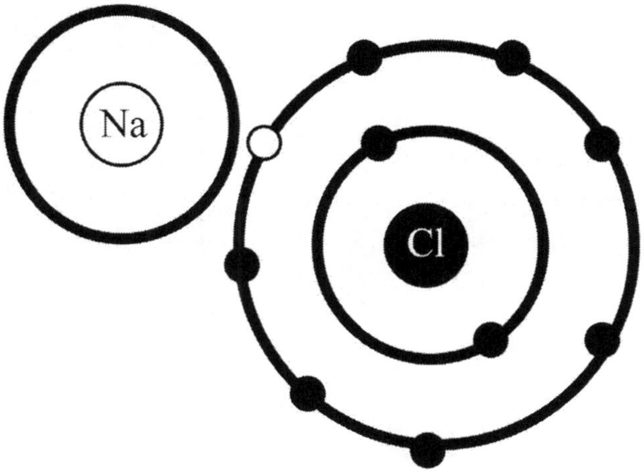

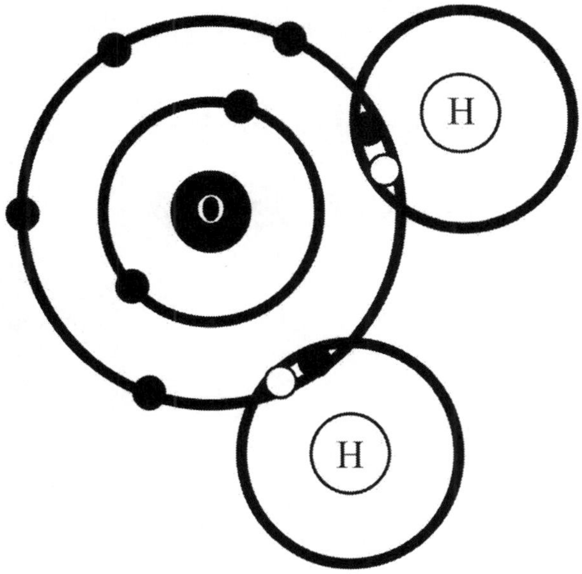

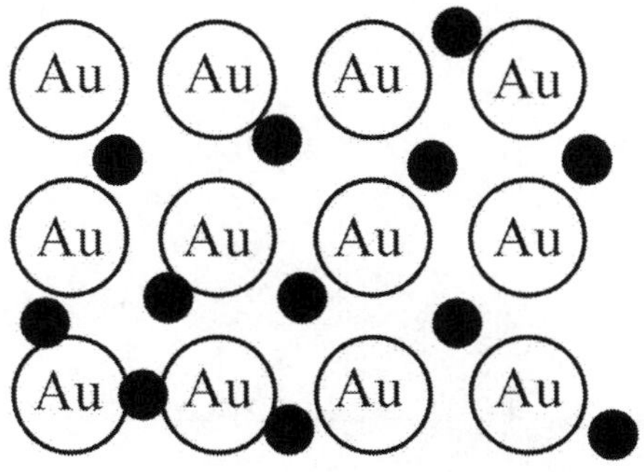

Experiment: Chocolate Bonding

Introduction

There are three main ways that atoms can chemically bond to form molecules. The first is called ionic bonding. In this type of bonding, one atom donates an electron to another to form the bond. The second way is called covalent bonding. In this type of bonding, one atom shares an electron with another atom to form the bond. In this experiment, you are going to use candy to demonstrate both these types of chemical bonding. The third type of bonding is metallic bonding, which we will discuss later.

Materials

______________________	______________________
______________________	______________________
______________________	______________________
______________________	______________________
______________________	______________________

Observations and Results

My Chocolate Version of Ionic Bonding

My Chocolate Version of Covalent Bonding

Conclusion

Written Assignment Lesson 1

Discussion Questions

1. What does all bonding involve?
2. What is valency?
3. What happens when an ionic bond forms?
4. How is a covalent bond formed?
5. What is metallic bonding?
6. What is a hydrogen bond?

Written Assignment Lesson 1

Student Assignment Sheet Lesson 2
Chemical Reactions – Part 1

Experiment: Exothermic or Endothermic?

Materials:
- ✓ Yeast
- ✓ Hydrogen peroxide
- ✓ Epsom salts
- ✓ Water
- ✓ 2 cups
- ✓ 2 thermometers

Procedure:
1. Read the introduction to this experiment and circle what you think the reactions will be.
2. Add 1 cup (240 mL) of hydrogen peroxide to one of the cups and 1 cup (240 mL) of warm water to the other. Place a thermometer in each up and wait 5 minutes. Record the temperature of both cups as the starting temperature.
3. Now, add 1 TBSP (8.5 g) of yeast to the hydrogen peroxide and gently stir. Then, add 1 TBSP (14.2 g) of Epsom salts to the water and gently stir. Wait until the temperature on the thermometer stabilizes and record the final temperature for the reaction.
4. Draw conclusions and complete your experiment sheet.

Vocabulary & Memory Work

- ☐ Vocabulary: chemical reaction, endothermic reaction, exothermic reaction, product, reactant
- ☐ Memory Work—This week, add the following elements to what you are working on memorizing:
  - ✓ 57-La-Lanthanum, 58-Ce-Cerium, 59-Pr-Praseodymium, 60-Nd-Neodymium
- ☐ Memory Work—Work on memorizing the Law of Conservation of Mass – In a chemical reaction, the mass of the products is equal to the mass of the reactants.

Sketch: Chemical Reaction

- Label the following: methane, oxygen, oxygen atoms, hydrogen atoms, carbon atom, water, carbon dioxide

Writing

- Reading Assignment: *DK Encyclopedia of Science* pg. 49 (Reactions) and pg. 52 (Chemical Reactions)
- Additional Research Readings: Chemical Reactions: *KSE* pg.162-163, *DK Chem* pg. 34-35, Energy and Chemical Reactions: *UIDS* pg. 146-147, What drives a reaction: *DK Chem* pg. 36-37

Dates

- 1620 – Francis Bacon publishes a book, *New Method*, in which he states that theories need to be supported by experiments.
- 1661 – Robert Boyle publishes *The Sceptical Chymist*, in which he says ideas should always be tested through experiments.
- 1789 – Antoine Lavoisier establishes that matter cannot be created or destroyed.

Sketch Lesson 2

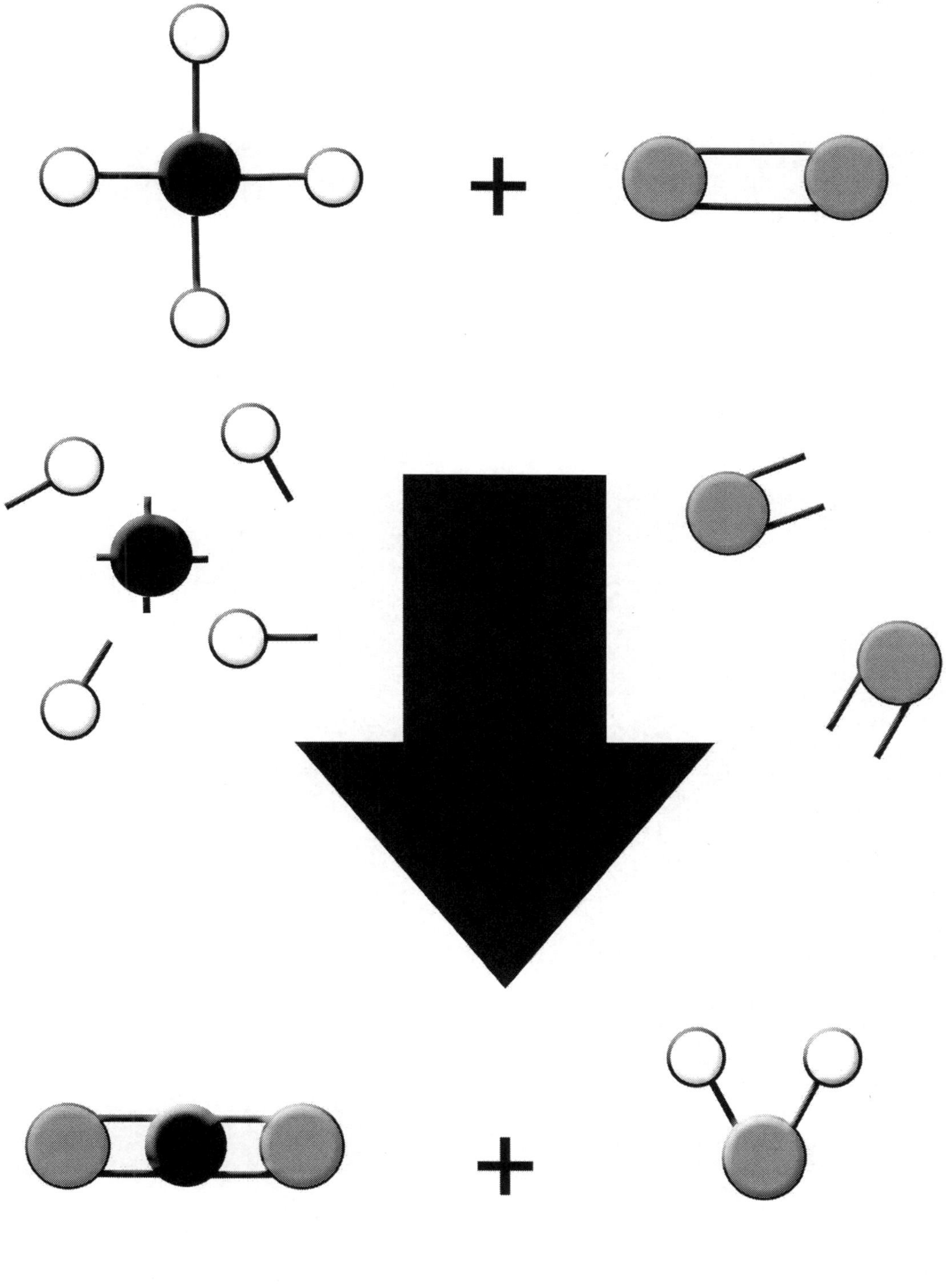

Experiment: Exothermic or Endothermic?

Introduction

In a chemical reaction, bonds are broken and reformed. This causes the reactants, or the substances you begin with, to be turned into new substances known as the products. Every chemical reaction involves energy that is either released or absorbed in the form of heat, light, or electricity. We call the heat-releasing reactions exothermic reactions and the heat-absorbing reactions are referred to as endothermic reactions. In this experiment, you are going to look at two different reactions to determine if they are exothermic or endothermic.

Hypothesis

Yeast + Hydrogen Peroxide =	Endothermic	Exothermic
Epsom Salts + Water =	Endothermic	Exothermic

Materials

_______________ _______________

_______________ _______________

_______________ _______________

_______________ _______________

_______________ _______________

Procedure

____

____

____

____

____

____

____

____

____

Observations and Results

Reaction	Endothermic or Exothermic
Yeast + Hydrogen Peroxide	
Epsom Salts + Water	

Conclusion

Written Assignment Lesson 2

Discussion Questions

Pg. 49, Reactions

1. What is a chemical change?
2. What is a physical change?

Pg. 52, Chemical Reactions

1. What sets the products of a chemical reaction apart from the reactants?
2. What happens in a chemical reaction and what does it require?
3. What is the difference between exothermic and endothermic reactions?
4. What is activation energy?

Written Assignment Lesson 2

Student Assignment Sheet Lesson 3
Reactivity

Experiment: How reactive is it?

Materials:

- ✓ Baking soda
- ✓ Chalk
- ✓ Iron nail (non-coated)
- ✓ Copper penny
- ✓ White vinegar
- ✓ 4 cups

Procedure:

1. Read the introduction to this experiment and write what you think the order the four reactants will be .
2. Pour ½ cup (120 mL) of vinegar into each of the 4 cups.
3. Add 1 tsp (3.5 g) of baking soda to the first cup. Observe and record how long it takes for a reaction to occur.
4. Repeat this process with 1 piece of chalk, 1 iron nail, and 1 copper penny in the other 3 cups. Observe and record how long it takes for a reaction to occur.
5. Draw conclusions and complete your experiment sheet.

Vocabulary & Memory Work

- ☐ Vocabulary: moles, reactivity
- ☐ Memory Work—This week, add the following elements to what you are working on memorizing:
  - ✓ 61-Pm-Promethium, 62-Sm-Samarium, 63-Eu-Europium, 64-Gd-Gadolinium
- ☐ Memory Work—Work on memorizing the Avogadro's number: 6.02×10^{23} and the Law of Conservation of Mass

Sketch: Anatomy of a Chemical Equation

- Label the following: reactants, products, numbers are used to balance the equation, symbols show what state the chemical is in, arrow to show the overall flow of the reaction

Writing

- Reading Assignment: *DK Encyclopedia of Science* pg. 53 (Describing Reactions) and pg. 66 (Reactivity Series)
- Additional Research Readings:
  - Reactivity: *UIDS* pp. 158-159

Dates

- 1811 – Italian scientist Amedeo Avogadro first proposes that the volume of a gas at a constant pressure and temperature is proportional to the number of atoms or molecules regardless of the nature of the gas.
- 1926 – Jean Perrin wins the Nobel Prize in Physics for determining the constant that Avogadro first proposed. He named the constant Avogadro's number in honor of the scientist.

Sketch Lesson 3

$$CaCl_{2(aq)} + 2AgNO_{3(aq)} \rightarrow Ca(NO_3)_{2(aq)} + 2AgCl_{(s)}$$

Experiment: How reactive is it?

Introduction

The reactivity series tells how reactive a given metal can be in its elemental state. Those at the top of the reactivity series, such as potassium, change quickly when exposed to air, while those at the bottom, such as gold, do not change at all. We can also use the periodic table to help us determine an element's propensity to react. For metals, reactivity decreases as you move from left to right on the periodic table and increases and you move down the table. In this experiment, you are going to test the reactivity of sodium (baking soda), calcium (chalk), iron (nail) and copper (penny) in vinegar.

Hypothesis

Write the four reactants (baking soda, chalk, nail, penny) in order of increasing reactivity:

____

____

Materials

_______________________ _______________________

_______________________ _______________________

_______________________ _______________________

_______________________ _______________________

_______________________ _______________________

Procedure

____

____

____

____

____

____

____

____

Observations and Results

	Baking soda + Vinegar	Chalk + Vinegar	Nail + Vinegar	Penny + Vinegar
Time to React*				

*Be sure to include your time units.

Conclusion

Written Assignment Lesson 3

Discussion Questions

Pg. 53, Describing Reactions

1. What does a chemical equation show?
2. How do chemists use the mole?
3. How many particles does 1 mole of a substance contain?

Pg. 66, Reactivity Series

1. What does the reactivity series tell us?
2. What is a displacement reaction?
3. What does it mean for steel to be galvanized and why is this done?

Written Assignment Lesson 3

Student Assignment Sheet Lesson 4
Chemical Reactions – Part 2

Experiment: Does the surface area affect cooking time?

> **CAUTION**
> **DO NOT DO THIS EXPERIMENT WITHOUT ADULT SUPERVISION!**
> Boiling hot water can cause severe damage. Be sure to use the proper safety gear.

Materials:
- ✓ 2 potatoes
- ✓ Pot
- ✓ Water
- ✓ Oven mitt
- ✓ Large Slotted Spoon

Procedure:
1. Read the introduction to this experiment and answer the question.
2. Take one potato and cut it in half. Then, cut one thin slice off of one of the halves and cut the other half in half. This will give you four sample potatoes, one full size potato, one half, one quarter and one thin slice.
3. Now, fill a pot three quarters of the way with water and place the four sample potatoes in it. Then, set the pot on a burner and turn the burner on medium high.
4. After the water has been boiling for 5 minutes, use the slotted spoon to remove each of the sample potatoes and check for doneness. If the sample is done, remove it from the pot and record how long it took to cook.
5. Repeat step 4 every 5 minutes until all four potato samples are fully cooked.
6. Draw conclusions and complete your experiment sheet.

Vocabulary & Memory Work
- ☐ Vocabulary: concentration, equilibrium, reversible reaction, irreversible reaction
- ☐ Memory Work—This week, add the following elements to what you are working on memorizing:
  - ✓ 65-Tb-Terbium, 66-Dy-Dysprosium, 67-Ho-Holmium, 68-Er-Erbium
- ☐ Memory Work—Work on memorizing Le Chatelier's Principle – If a change is made to a reaction in equilibrium, the reaction will adjust itself to counter the effects of that change.

Sketch: Le Chatelier's Principle
- Label the following: a reaction in equilibrium, the addition of more reactants causes the forward reaction to speed up, the addition of more products causes the reverse reaction to speed up

Writing
- Reading Assignment: *DK Encyclopedia of Science* pg. 54 (Reversible Reactions) and pg. 55 (Rates of Reactions)
- Additional Research Readings:
  - Reversible Reactions: *UIDS* pp. 162-163
  - Rates of Reactions: *DK Chem* pp.38-39

Dates
- 1850-1936 – Henry Louis Le Chatelier lives. He is famous for the equilibrium law that he discovers during these years.

Sketch Lesson 4

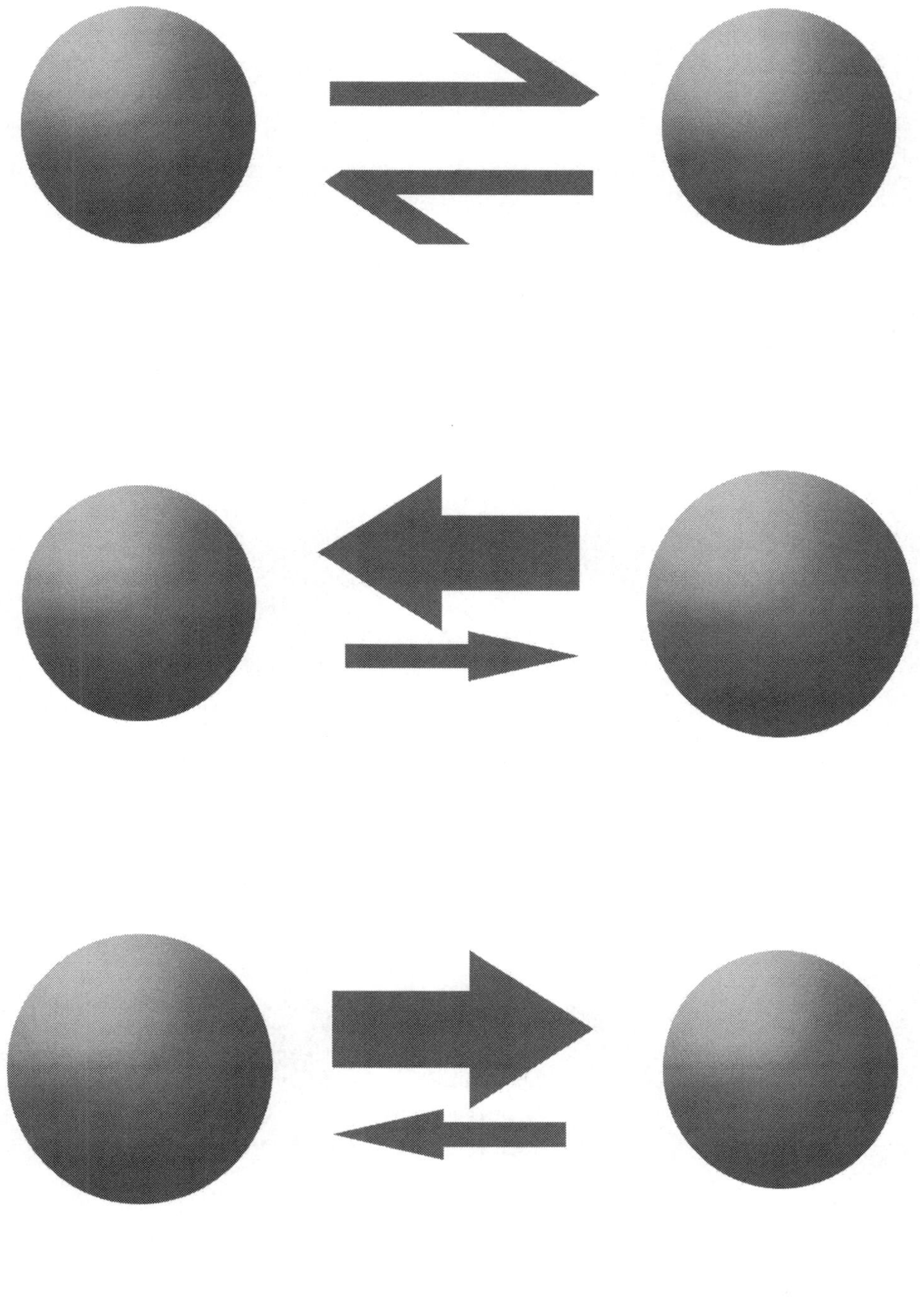

Experiment: Does the surface area affect cooking time?

Introduction

The collision theory states that for a chemical reaction to occur the particles involved must hit one another with enough force or energy to break the bonds involved. In other words, there must collide with enough energy to overcome the activation energy required for the reaction to occur. If they don't, they will simply bounce off one another and no reaction will take place. Many factors such as temperature, concentration, and surface area can affect the rate at which the molecules collide in a reaction. In this experiment, you are going to examine how surface area can affect the rate of a reaction.

Hypothesis

____

____

____

____

Materials

________________________ ________________________

________________________ ________________________

________________________ ________________________

________________________ ________________________

________________________ ________________________

Procedure

____

____

____

____

____

____

____

____

Observations and Results

	Potato Slice	Potato Quarter	Potato Half	Whole Potato
Time to Cook*				

*Be sure to include your time units.

Conclusion

Written Assignment Lesson 4

Discussion Questions

pg. 54, Reversible Reactions

1. What is the difference between an irreversible and a reversible reaction?
2. What happens when a reversible reaction reaches equilibrium?
3. How does a chemical clock work?

pg. 55, Rates of Reactions

1. How does temperature typically affect the rate of a reaction? Light? Concentration? Pressure?
2. Briefly explain collision theory.

Written Assignment Lesson 4

Student Assignment Sheet Lesson 5
Catalysts

Experiment: Can I speed up the reaction?

Materials:

- ✓ Carbonated water
- ✓ Sugar
- ✓ 2 cups

Procedure:

1. Read the introduction to this experiment and write down which reaction you think will go faster.
2. Label your two cups cup #1 and cup #2.
3. Pour one cup (240 mL) of carbonated water into each one of the cups. Wait till you see bubbles begin to form on the sides of the cup and record the time as your experiment start time.
4. Now, add 1 tsp (4 g) of sugar to cup #2. Record your observations about the reaction that is happening in both cups. When the reactions reach equilibrium (*i.e. no more bubbles are coming up*), record the end time for each cup in the experiment.
5. Draw conclusions and complete your experiment sheet.

Vocabulary & Memory Work

- ☐ Vocabulary: catalyst, enzyme
- ☐ Memory Work—This week, add the following elements to what you are working on memorizing:
  - ✓ 69-Tm-Thulium
  - ✓ 70-Yb-Ytterbium
  - ✓ 71-Lu-Lutetium
  - ✓ 72-Hf-Hafnium

Sketch: Reaction Pathway

- Label the following: reactants, products, normal path of a reaction, activation energy, reaction path with the addition of a catalyst, activation energy when a catalyst has been introduced

Writing

- Reading Assignment: *DK Encyclopedia of Science* pp. 56-57 (Catalysts)
- Additional Research Readings:
  - Catalysts: *KSE* pg. 176
  - Enzymes: *KSE* pg. 177
  - Rates of Reactions: *UIDS* pp. 160-161

Dates

- 1767 – Joseph Priestley inventes carbonated water when he discovers that he can infuse water with carbon dioxide by suspending a bowl of water above a beer.
- 1909 – Wilhelm Ostwald is awarded the Nobel Prize for Chemistry for his work with catalysts.

Sketch Lesson 5

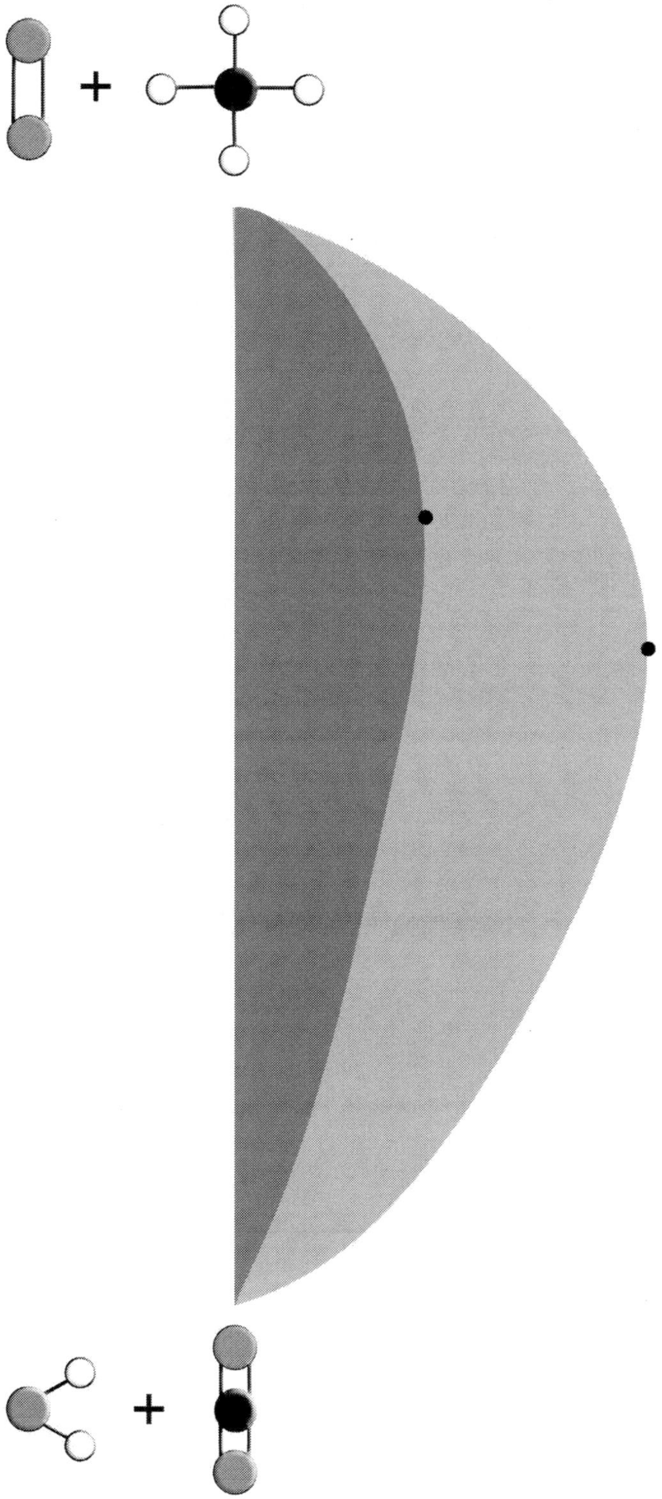

Experiment: Can I speed up the reaction?

Introduction

The rate of a reaction can be changed by varying the temperature, pressure or concentration of the reactants involved. The rate can also be modified through the addition of a catalyst. A catalyst will alter the speed of a reaction, but it will not be changed by the process. In this experiment, you are going to look at how the addition of a catalyst can adjust the rate of a reaction.

Hypothesis

____

____

____

____

Materials

____________________ ____________________

____________________ ____________________

____________________ ____________________

____________________ ____________________

____________________ ____________________

Procedure

____

____

____

____

____

____

____

____

____

Observations and Results

	Cup #1	Cup #2
Time to React*		

*Be sure to include your time units.

Conclusion

Written Assignment Lesson 5

Discussion Questions

1. Why are catalysts added to a reaction?
2. How can catalysts speed up a reaction?
3. What are zeolites?
4. What are enzymes and how does our body use them?
5. What are some ways that catalysts are used?

Written Assignment Lesson 5

Student Assignment Sheet Lesson 6
Oxidation and Reduction

Experiment: What happens when iron is oxidized?

Materials:
- ✓ Steel wool
- ✓ Vinegar
- ✓ Jar with lid
- ✓ Ammonia
- ✓ Hydrogen peroxide
- ✓ 2 bowls

Procedure:
1. Read the introduction to this experiment and answer the question.
2. The day before you do this experiment, prepare the iron acetate solution by placing the steel wool in one of the jars and covering it with vinegar. Put the lid on the jar and let the steel wool soak for 24 hours.
3. The next day, put on your safety googles before adding 2 TBSP (30 mL) of the iron acetate solution you made to the second jar. Then, add 2 to 4 TBSP (30 mL) of ammonia (until you see a color change that remains) and record your observations.
4. Next, add 1 TBSP (30 mL) of hydrogen peroxide and record your observations.
5. Draw conclusions and complete your experiment sheet.

Vocabulary & Memory Work
- ☐ Vocabulary: oxidation, reduction, redox reaction
- ☐ Memory Work—This week, add the following elements to what you are working on memorizing:
  - ✓ 73-Ta-Tantalum
  - ✓ 74-W-Tungsten
  - ✓ 75-Re-Rhenium
  - ✓ 76-Os-Osmium

Sketch: Oxidation/Reduction Reaction
- Label the following: the oxidizing agent gives up oxygen to another molecule, the reducing agent gives hydrogen to another molecule, the reduced molecule has gained a hydrogen atom, the oxidized molecule has gained an oxygen atom

Writing
- Reading Assignment: *DK Encyclopedia of Science* pp. 64-65 (Oxidation and Reduction)
- Additional Research Readings:
  - Oxidation and Reduction: *KSE* pg. 178, *UIDS* pp. 148-149, *DK Chem* pp. 40-41

Dates
- No dates to be entered this week.

Sketch Lesson 6

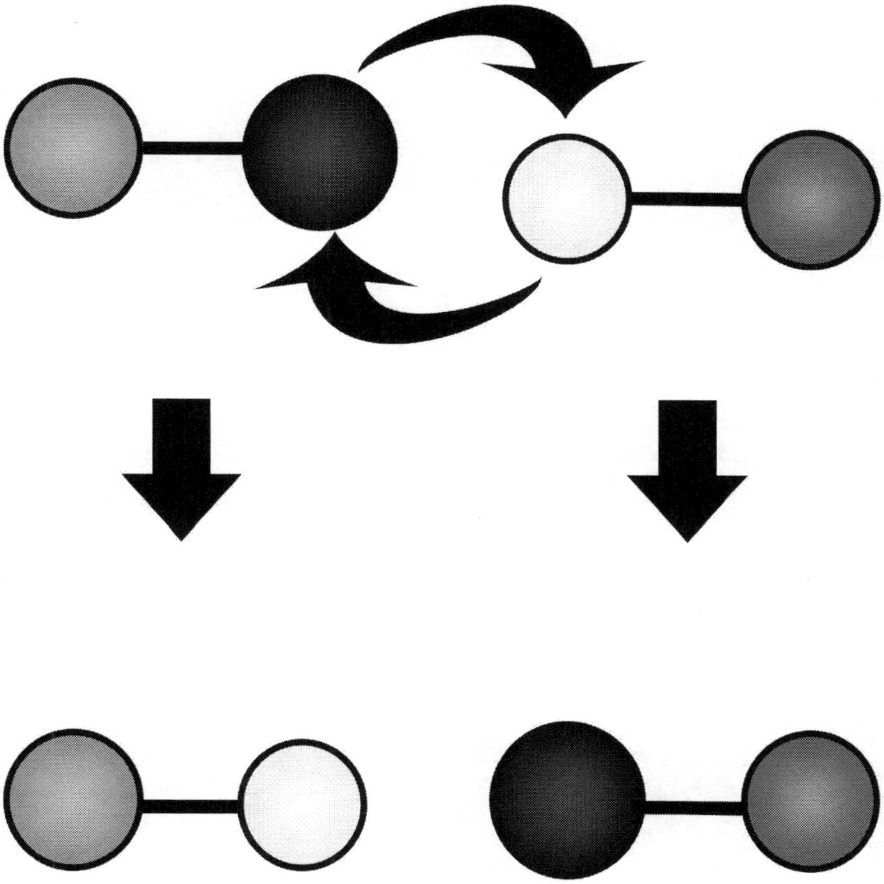

Experiment: What happens when iron is oxidized?

Introduction

Combustion, rusting and respiration are all examples of oxidation and reduction reactions. These reactions typically involve the movement of oxygen and hydrogen atoms, however an oxidation/reduction reaction can occur even when oxygen or hydrogen are not present. The key is to have the movement of electrons present in the reaction. In other words, an atom is oxidized when it loses an electron and reduced when it gains one. In this experiment, you are going to look at what happens to iron atoms when they are oxidized.

Hypothesis

____

____

____

Materials

______________ ______________

______________ ______________

______________ ______________

______________ ______________

______________ ______________

Procedure

____

____

____

____

____

____

____

____

____

Observations and Results

Conclusion

Written Assignment Lesson 6

Discussion Questions

1. Why does oxidation and reduction always occur together?
2. What are some common oxidation/reduction reactions?
3. What happens to a substance when it is reduced?
4. What happens to a substance when it is oxidized?
5. What happens to electrons in an oxidation and reduction reaction?

Written Assignment Lesson 6

Unit 7

Acids and Bases

Unit 7: Acids and Bases
Vocabulary Sheet

Define the following terms as they are assigned on your Student Assignment Sheet.

1. Acid – ______________________________

2. Acidic Solution – ______________________________

3. Dissociation – ______________________________

4. Alkali – ______________________________

5. Alkaline Solution – ______________________________

6. Base – ______________________________

7. Indicator – ______________________________

8. Buffer – ______________________________

9. pH – ____

____

____

10. Titration – ____

____

____

11. Neutralization – ____________________________________

____

____

12. Salt – ____

____

____

Student Assignment Sheet Lesson 1
Acids

Experiment: Can I change an acid?

Materials:

- ✓ Cranberry juice
- ✓ Lemon juice
- ✓ Baking soda
- ✓ Clear cup

Procedure:

1. Read the introduction to this experiment and then write down what you think will happen when you add the baking soda and lemon juice to the cranberry juice.
2. Begin by pouring 1 cup (240 mL) of cranberry juice into the cup. Then, add 1 TBSP (15 mL) of lemon juice. Note any changes and record the observations.
3. Then, add 1 TBSP (14 g) of baking soda. Note any changes and record the observations.
4. Finally, add another TBSP (15 mL) of lemon juice. Note any changes and record the observations.
5. Draw conclusions and complete the experiment sheet.

Vocabulary & Memory Work

- ☐ Vocabulary: acid, acidic solution, dissociation
- ☐ Memory Work—This week, add the following elements to what you are working on memorizing:
  - ✓ 77-Ir-Iridium, 78-Pt-Platinum, 79-Au-Gold, 80-Hg-Mercury

Sketch Assignment: Strong Acids vs. Weak Acids

- Label the following: strong acid, weak acid, positive hydrogen ions (in each box), negative ions (in each box), weak acid compounds

Writing

- Reading Assignment: *DK Encyclopedia of Science* pp. 68-69 (Acids)
- Additional Research Readings:
  - Acids: *KSE* pg. 184
  - Acids: *UIDS* pg. 150

Dates

- 11th century – Arabic chemists discover how to make acidic compounds such as sulfuric, nitric and hydrochloric acids.
- 1675 – Robert Boyle, an Irish chemist, suggests that acids might be made up of sharp pointed particles.
- 1770's – Antoine Lavoisier, a French chemist, suggests that acids form oxygen compounds when dissolved in water, so they must all contain oxygen.
- 1887 – Svante Arrhenius, a Swedish chemist, develops the modern theory on acids. He states that all acids contain hydrogen ions, which give these compounds their unique properties.

Sketch Lesson 1

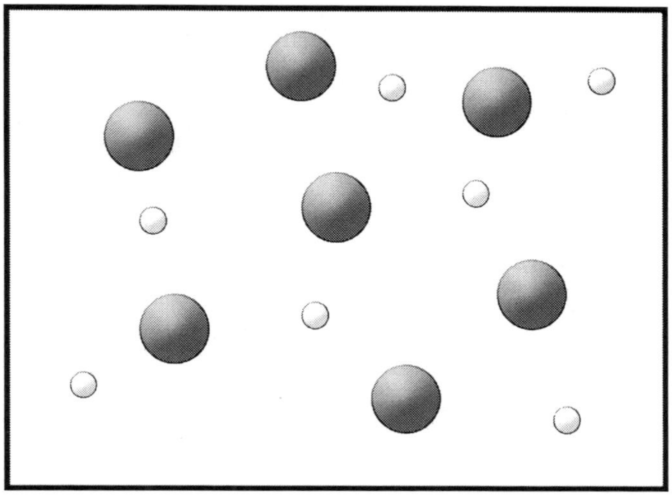

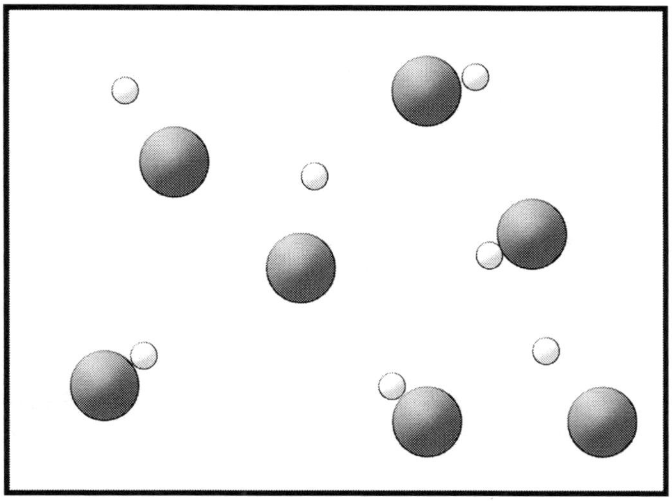

Experiment: Can I change an acid?

Introduction

Acids release positive hydrogen ions when in a solution. They are in batteries, in our stomachs and in the food we eat and drink. Acids are responsible for the sour taste that you find in food. Some of the fruits we eat, such as pears and blueberries, contain pigments that change colors in the presence of an acid or base. The pigment anthocyanin, which is found in cranberries, is red when in the presence of an acid and blue in the presence of a base. In this experiment, you are going to try to alter the color of acidic cranberry juice with other common kitchen substances.

Hypothesis

____

____

____

Materials

________________ ________________

________________ ________________

________________ ________________

________________ ________________

________________ ________________

Procedure

____

____

____

____

____

____

____

____

____

Observations and Results

	Color of the Solution
Cranberry Juice	
Cranberry Juice + 1TBSP Lemon Juice	
Cranberry Juice + 1 TBSP Lemon Juice + 1TBSP Baking Soda	
Cranberry Juice + 1 TBSP Lemon Juice + 1TBSP Baking Soda + 1 TBSP Lemon Juice	

Conclusion

Written Assignment Lesson 1

Discussion Questions

1. What happens to acids in water?
2. What is the difference between strong acids and weak acids?
3. Why are strong acids used in batteries?
4. Why is acid rain so damaging to the environment?
5. What do acids always do to carbonates?

Written Assignment Lesson 1

Student Assignment Sheet Lesson 2
Bases

Experiment: Is it an acid or a base?

Materials:

- ✓ 6 cups
- ✓ Red cabbage solution
- ✓ Water
- ✓ Vinegar
- ✓ Baking soda
- ✓ Sprite
- ✓ Ammonia
- ✓ Lemon Juice
- ✓ Eye dropper

Procedure:

How to make the red cabbage solution *– The night before you do this experiment you will need to make the red cabbage solution. Have your parent or teacher cut up a head of cabbage, place it in a pot and put it on the stove. Turn the heat to high and boil the cabbage for 15 minutes. Remove the cooked cabbage and let the solution cool. Once it is cool put it in a container and store it in a glass jar in the refrigerator. You will also need this solution for the experiment in Lesson 4.*

1. Read the introduction to this experiment and write down whether you think vinegar, baking soda, sprite, ammonia and lemon juice are acids or bases.
2. Label the cups #1 through #6. Pour ¼ cup (60 mL) of water into cup #1, a ¼ cup (60 mL) of vinegar into cup #2, ¼ cup (60 mL) of Sprite into cup #3, ¼ cup (60 mL) of ammonia into cup #5, and ¼ cup (60 mL) of lemon juice into cup #6. Add ¼ cup (60 mL) of water to cup #4 and dissolve 1 TBSP (14 g) of baking soda in it.
3. Add 1 tsp (5 mL) of red cabbage juice to each cup, mix thoroughly. Be sure to either thoroughly rinse the spoon after each one or use a different spoon for each cup.
4. Compare the color of the solution in each cup to the color of the original solution. If the solution is the same color as the original, the substance is neutral. If the solution turned to a light red-purple, the substance was a weak acid. If the solution was red, the substance was a strong acid. If the solution turned blue, the substance was a weak base. If the solution was blue-green, the substance was a strong base.
5. Record the color of the solution on the results chart of the experiment sheet and then decide if the solutions were acidic, neutral or basic.
6. Draw conclusions and complete the experiment sheet.

Vocabulary & Memory Work

- ☐ Vocabulary: alkali, alkaline solution, base, indicator
- ☐ Memory Work—This week, add the following elements to what you are working on memorizing:
  - ✓ 81-Tl-Thallium, 82-Pb-Lead, 83-Bi-Bismuth, 84-Po-Polonium

Sketch: Strong Bases vs. Weak Bases

- Label the following: strong base, weak base, negative hydroxide ions (in each box), positive ions (in each box), weak base compounds

Writing

- Reading Assignment: *DK Encyclopedia of Science* pp. 70-71 (Alkalis and Bases)
- Additional Research Readings: Bases and Alkalis: *KSE* pg. 185, Bases: *UIDS* pg. 151

Dates

- 1754 – French chemist Guillaume-François Rouelle states that a base is a substance that reacts with an acid to produce a solid (or salt).

Sketch Lesson 2

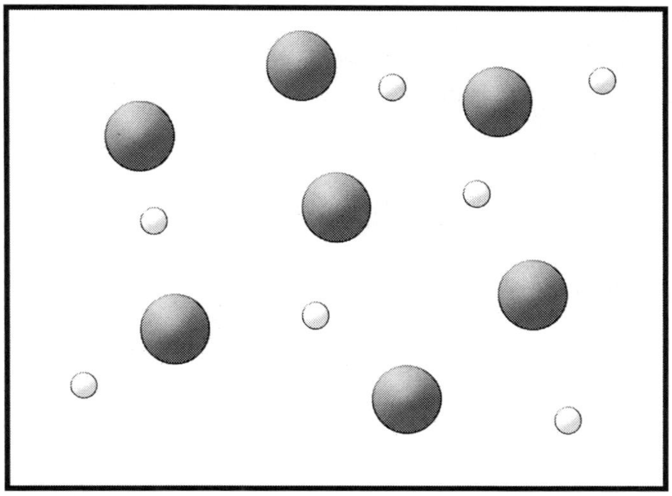

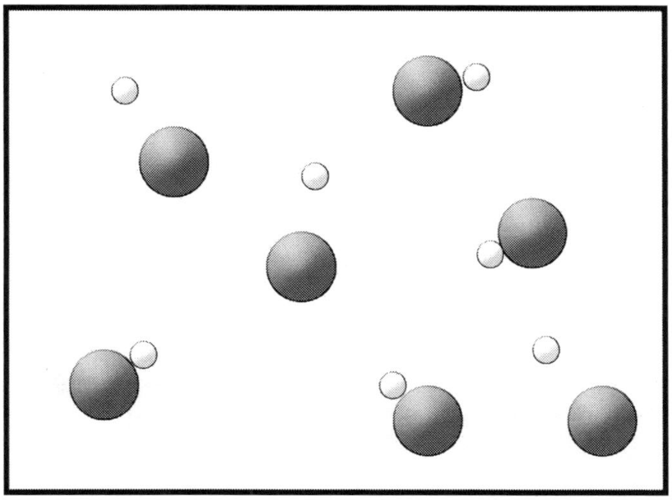

Experiment: Is it an acid or a base?

Introduction

The solutions you see in the everyday life all have either acidic or basic properties. In other words, they can either release positive hydrogen ions into a solution (acid) or they can release negative hydroxide ions into a solution (base). In the laboratory, scientists can use indicators to tell whether or not a solution is acidic or basic. This indicator will change color depending upon the properties of the solution. In this experiment, you are going use an indicator to look at several different common household substances to determine whether they are acids or bases.

Hypothesis

Acids: ____

Bases: ____

Materials

____________________ ____________________

____________________ ____________________

____________________ ____________________

____________________ ____________________

____________________ ____________________

Procedure

____

____

____

____

____

____

____

____

____

Observations and Results

Substance	Color	Acid or Base
Water		
Vinegar		
Baking soda		
Sprite		
Ammonia		
Lemon juice		

Conclusion

Written Assignment Lesson 2

Discussion Questions

1. What do bases do to acids?
2. What do all alkalis have in common?
3. Where does the base calcium carbonate come from and what is it used for?
4. What is the difference between a strong base and a weak base?
5. What is neutralization?

Written Assignment Lesson 2

Student Assignment Sheet Lesson 3
Measuring Acidity (pH)

Experiment: Measuring pH

Materials:

- ✓ Lemon
- ✓ Tomato
- ✓ Saliva
- ✓ Milk
- ✓ Bleach
- ✓ Toothpaste
- ✓ Dish soap
- ✓ pH paper
- ✓ Gloves

pH Scale

- ☞ pH 0 to just under 7 is an acid.
- ☞ pH 7 is neutral.
- ☞ pH just over 7 to 14 is a base.

Procedure:

1. Read the introduction to this experiment.
2. Put on a pair of gloves and lay out each of your test substances in a row. Touch the pH paper to the surface of the test substance and wait for the paper to change color.
3. Compare the color of the paper with the scale included with your pH paper and record the approximate pH of the substance. Then, use the number scale above to determine if the substance is an acid, base or neutral.
4. Draw conclusions and complete the experiment sheet.

Vocabulary & Memory Work

- ☐ Vocabulary: buffer, pH
- ☐ Memory Work—This week, add the following elements to what you are working on memorizing:
  - ✓ 85-As-Astatine, 86-Rn-Radon, 87-Fr-Francium, 88-Ra-Radium

Sketch: pH scale

- Label the following: the pH scale from 0 to 14, acidic range, neutral, basic range
- Then, use varying shades of red to show that the strength of an acid increases as you go down the scale and varying shades of blue to show that the strength of a base increases as you go up.

Writing

- Reading Assignment: *DK Encyclopedia of Science* pg. 72 (Measuring Acidity)
- Additional Research Readings:
  - Indicators and pH: *KSE* pg. 186
  - Acids and Bases Strength: *UIDS* pg. 152
  - Acids and Bases: *DK Chem* pg. 42-43

Dates

- 1909 – Danish chemist Soren Sorenson comes up with a logarithmic scale to show the concentration of the hydrogen ions in a solution, which ranged from 0 to 14. Today this scale is known as the pH scale.

Sketch Lesson 3

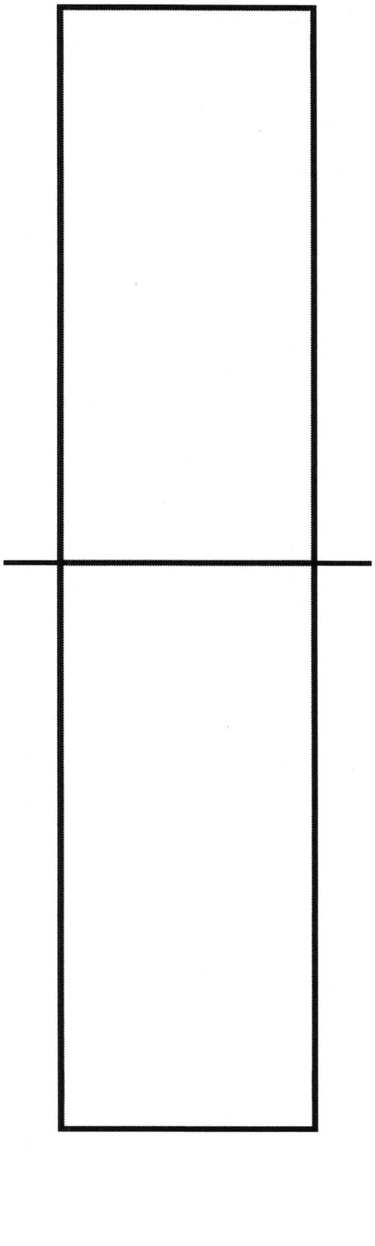

Experiment: Measuring pH

Introduction

The pH of a solution is a measure of the acidity or alkalinity of the solution. The pH scale goes from zero to fourteen. A solution with a pH below seven is considered an acid and a solution with a pH above seven is considered a base. If the solution has a pH of seven, it is considered to be neutral. The closer an acidic solution is to having a pH of zero, the stronger it is, and the stronger an alkaline solution is to having a pH of fourteen the stronger it is. In this experiment, you are going to objectively measure the pH of several common household substances.

Materials

______________________ ______________________

______________________ ______________________

______________________ ______________________

______________________ ______________________

______________________ ______________________

Procedure

____

____

____

____

____

____

____

____

____

Observations and Results

Substance	pH	acid, base or neutral
Lemon		

Substance	pH	acid, base or neutral
Tomato		
Saliva		
Milk		
Bleach		
Toothpaste		
Liquid dish soap		

Conclusion

Written Assignment Lesson 3

Discussion Questions

1. What does pH stand for and what does it measure?
2. What is the pH scale based on?
3. Why do chemists use indicators?

Written Assignment Lesson 3

Student Assignment Sheet Lesson 4
Neutralization and Salts

Experiment: Can I neutralize an acid?

Materials:

- ✓ Vinegar
- ✓ Ammonia
- ✓ Red cabbage solution (*Use the solution you made in Lesson 2.*)
- ✓ Water
- ✓ Safety glasses
- ✓ 4 cups

Procedure:

1. Read the introduction to this experiment and answer the question.
2. Make the solutions using the following directions:
   - ☑ **Acidic Solution** – Mix ¼ cup (60 mL) of vinegar with ⅓ cup (79 mL) of water and label the cup "Acid".
   - ☑ **Basic Solution** – Mix ¼ cup (60 mL) of ammonia with 2 TBSP (30 mL) of water and label the cup "Base".
3. Next, mix ¼ cup (60 mL) of the acidic solution with ¼ cup (60 mL) of the red cabbage solution, which will act as an indicator and label the cup "test". In another cup, mix ¼ cup (60 mL) of water with ¼ cup (60 mL) of the red cabbage solution and label the cup "control". (***Note****—The color of the test solution should be pink and the color of the control solution should be purple-blue.*)
4. Next, slowly add the basic solution 1 tsp (5 mL) at a time into the test solution until you achieve the color of the neutral control solution.
5. Record how many drops it took to neutralize the test solution, draw conclusions and complete the experiment sheet.

Vocabulary & Memory Work

- ☐ Vocabulary: neutralization, salts, titration
- ☐ Memory Work—This week, add the following elements to what you are working on memorizing:
  - ✓ 89-Ac-Actinium, 90-Th-Thorium, 91-Pa-Protactinium, 92-U-Uranium

Sketch: Acid/Base Equation

- Label the following: acid, hydrochloric acid, base, sodium hydroxide, salt, sodium chloride, water

Writing

- Reading Assignment: *DK Encyclopedia of Science* pg. 73 (Salts)
- Additional Research Readings:
  - Forming Salts: *DK Chem* pg. 44-45, Salts: *UIDS* pp. 153-154

Dates

- 1923 – Danish chemist Johannes Bronsted and English chemist Martin Lowry both suggest a change to Arrhenius' definition of acids and bases. They define an acid as a chemical that donates hydrogen ions (or protons) and a base as one that accepts them.

Sketch Lesson 4

$$HCl_{(aq)} + NaOH_{(aq)} \rightarrow NaCl_{(aq)} + H_2O_{(l)}$$

__________ + __________ → __________ + __________

Experiment: Can I neutralize an acid?

Introduction

Neutralization is the process by which a solution is made neither acidic nor alkaline. Generally, a scientist will use titration to achieve the neutralization of an acid or a base. For this method, he or she will use an indicator solution that changes color with changing pH. This way the scientist will know when he or she has come close to a neutral pH of 7. In this experiment, you will use titration to attempt to reach a neutral pH by combining an acidic solution with a basic solution.

Hypothesis

I think that I will see____

Materials

Procedure

Observations and Results

Conclusion

Written Assignment Lesson 4

Discussion Questions

1. What are salts?
2. Name a few examples of salts.
3. How is a salt made?
4. How do the strong ionic bonds affect the boiling and melting points of salts?

Written Assignment Lesson 4

Chemistry Unit 8

Chemistry of Life

Unit 8: Chemistry of Life
Vocabulary Sheet

Define the following terms as they are assigned on your Student Assignment Sheet.

1. Isomer – ______________________________

____

____

2. Monomers – ______________________________

____

____

3. Organic Compound – ______________________________

____

____

4. Polymer – ______________________________

____

____

5. Anabolism – ______________________________

____

____

6. Catabolism – ______________________________

____

____

7. Carbohydrates – ______________________________

____

____

8. Lipids – ______________________________

____

____

9. Minerals – ____

____

____

10. Protein – ____

____

____

11. Starch – ____

____

____

12. Vitamins – ____

____

____

13. Alcohol – ____

____

____

14. Fermentation – ____

____

____

15. Yeast – ____

____

____

Student Assignment Sheet Lesson 1
Organic Chemistry

Experiment: Is there carbon in sugar or salt?

Materials:
- ✓ Sugar
- ✓ Salt
- ✓ Candle
- ✓ 2 metal spoons
- ✓ Hot mitt

CAUTION

DO NOT DO THIS EXPERIMENT WITHOUT ADULT SUPERVISION!

This experiment involves an open flame. Be sure to use the proper safety gear.

Procedure:
1. Read the introduction to this experiment and make a hypothesis.
2. After putting on safety goggles and protective gloves, light your candle and let it burn for a few minutes.
3. Meanwhile, add about ½ tsp (2.5 g) of sugar to one of the spoons. Then, using a pair of tongs, hold the end of the spoon with the sugar over the flame for 7 minutes. Observe what happens and record what you see.
4. Then, add about ½ tsp (2.8 g) of salt to the other spoon. Using a pair of tongs, hold the end of the spoon with the salt over the flame for 7 minutes. Observe what happens and record what you see.
5. Draw conclusions and complete the experiment sheet.

Vocabulary & Memory Work
- ☐ Vocabulary: isomer, monomers, organic compounds, polymer
- ☐ Memory Work—This week, work on memorizing the carbon cycle (*see the sketch for details*) and add the following elements to what you are working on memorizing:
  - ✓ 93-Np-Neptunium, 94-Pu-Plutonium, 95-Am-Americium, 96-Cm-Curium

Sketch: Carbon Cycle
- Label the following: carbon dioxide in the air, organic compounds from plants decompose into other organic compounds and carbon dioxide, plants make sugars from the carbon dioxide in the air, animals get organic compounds from eating plants, organic compounds in plants and fuels release carbon dioxide into the air as they burn, organic compounds in animals turn into organic compounds and carbon dioxide as the animals breath

Writing
- Reading Assignment: *DK Encyclopedia of Science* pg. 40 (Carbon) and pg. 41 (Organic Chemistry)
- Additional Research Readings:
  - Organic Chemistry: *KSE* pp. 174-175, *UIDS* pp. 190-191
  - Chemistry of Life: *DK Chem* pp. 50-51

Dates
- 1808 – Jons Berzelius, a Swedish chemist, first uses the term organic chemistry to refer to the chemistry of living things.
- 1828 – The meaning of organic chemistry changes to refer to the chemistry of carbon when Friedrich Wohler succeeds in synthesizing a natural carbon compound in the lab.

Sketch Lesson 1

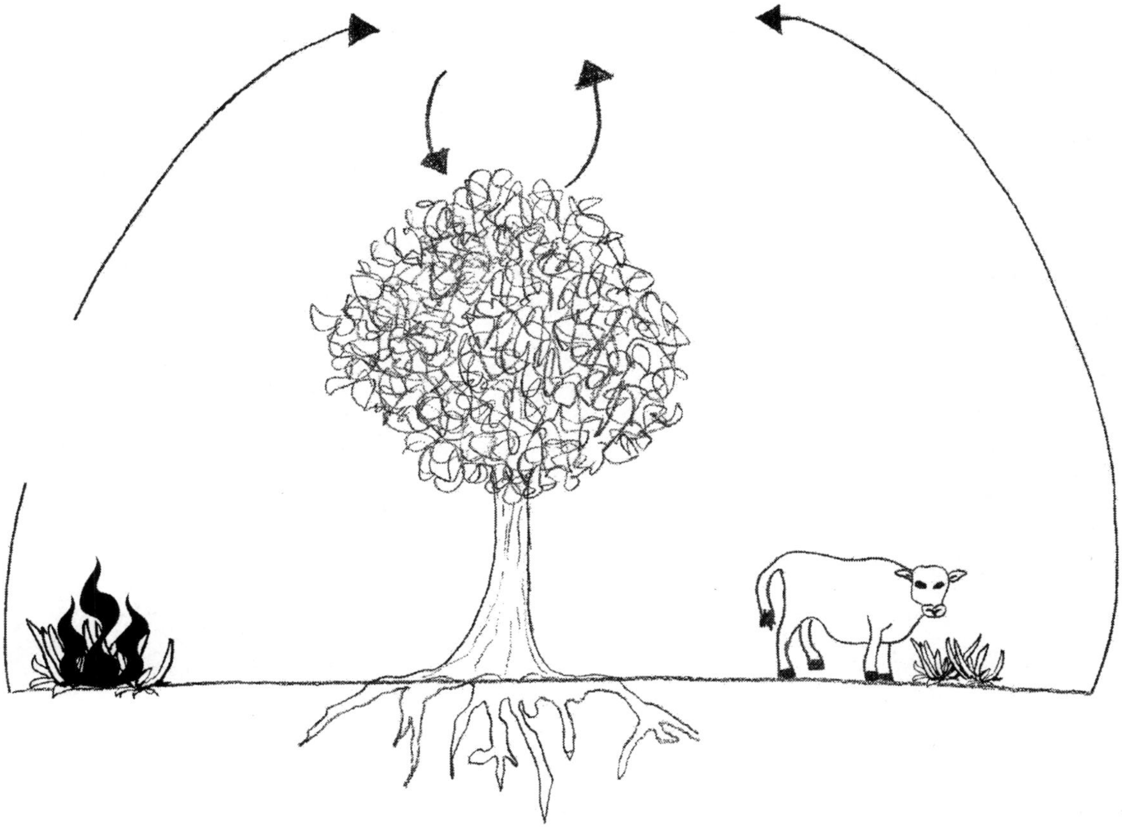

Experiment: Is there carbon in sugar or salt?

Introduction

Organic compounds are substances that contain carbon. They generally have low melting and boiling points, which means that these compounds burn fairly easily in the air and always give off carbon dioxide. If there is incomplete combustion, which normally occurs in a non-laboratory environment, some elemental carbon or soot will remain. On the other hand, inorganic compounds typically have high melting and boiling points, which means that they require a fair amount of heat energy before they will burn. In this week's experiment, you are going to use heat to test two different common household chemicals for the presence of carbon.

Hypothesis

____

____

____

Materials

______________	______________
______________	______________
______________	______________
______________	______________
______________	______________

Procedure

____

____

____

____

____

____

____

Observations and Results

Substance	Was Carbon Present?
sugar	
salt	

Conclusion

Written Assignment Lesson 1

Discussion Questions

Pg. 40, Carbon

1. Where is carbon found in nature?
2. What is the greenhouse effect?

Pg. 41, Organic Chemistry

1. How is carbon different from other elements?
2. What are the three main groups of organic compounds?
3. What are plastics made from?

Written Assignment Lesson 1

Student Assignment Sheet Lesson 2
Chemistry of the Human Body

Experiment: Does saliva digest food?

Materials:

- ✓ 2 slices of bread
- ✓ Water
- ✓ Saliva
- ✓ 2 plastic bags

Procedure:

1. Read the introduction to this experiment and make a hypothesis.
2. Label the bags #1 and #2 and then place a slice of bread into each of the bags.
3. Add ¼ cup (60 mL) of water to bag #1, seal it and set it on the counter where it won't be disturbed.
4. Spit into a cup several times, until you have about 2 TBSP (30 mL) of saliva. Add 2 TBSP (30 mL) of water and mix well. Pour the mixture over the bread in bag #2, seal the bag and set in on the counter next to bag #1.
5. Check the bags every 15 minutes over 4 hours and record what you see happening. Each time you check on the bread, be sure to quantify the amount of bread that has been broken down. For example, you could assign a percentage or use a scale, e.g. 1 (*not broken down*) to 10 (*completely broken down*).
6. After two hours, draw conclusions and complete the experiment sheet.

Vocabulary & Memory Work

- ☐ Vocabulary: anabolism, catabolism
- ☐ Memory Work—This week, continue to work on memorizing the carbon cycle and add the following elements to what you are working on memorizing:
  - ✓ 97-Bk-Berklium, 98-Cf-Californium, 99-Es-Einsteinium, 100-Fm-Fermium

Sketch: Two Types of Respiration

- Label the following: Aerobic Respiration: glucose, oxygen, carbon dioxide, water, energy; Anaerobic Respiration: glucose, lactic acid, energy

Writing

- Reading Assignment: *DK Encyclopedia of Science* pp. 76-77 (Chemistry of the Human Body)
- Additional Research Readings:
  - Enzymes: *KSE* pg. 177, Metabolism: *UIDS* pg. 332-333

Dates

- 1833 – The first enzyme diastase, which we call amylase today, is discovered by Anselme Payen.
- 1950 to today – The field of biochemistry, the study of the chemistry within living things, has greatly advanced with the development of new technology such as chromatography, X-ray diffraction, NMR spectroscopy, radio-isotopic labeling, electron microscopy, and molecular dynamics' simulations.

Sketch Lesson 2

$$C_6H_{12}O_6 + 6O_2 \rightarrow 6CO_2 + 6H_2O + 2900 \text{ kJ}$$

$$C_6H_{12}O_6 \rightarrow 2C_3H_6O_3 + 120 \text{ kJ}$$

Experiment: Does saliva digest food?

Introduction

Digestion is a collection of chemical processes that occur in the body. Through these processes the food you eat is broken down into the organic compounds that your body needs. Digestion is performed within several of the organs found in your body and it uses different enzymes as catalysts to speed up the action along the way. In this week's experiment, you are going to look at one of your body's fluids to see if it aides in digesting your food.

Hypothesis

___

___

___

Materials

___ ___

___ ___

___ ___

___ ___

Procedure

___

___

___

___

___

___

___

___

___

___

Observations and Results

Liquid	Amount Digested
water	
water + saliva	

Conclusion

Written Assignment Lesson 2

Discussion Questions

1. How does the body make glucose?
2. Explain the chemistry behind digestion.
3. What is the difference between catabolic and anabolic reactions?
4. What is the purpose of respiration and what are the two types of respiration?

Written Assignment Lesson 2

Student Assignment Sheet Lesson 3
Chemistry of Food

Experiment: Is it a sugar or a starch?

Materials:
- ✓ Benedict's solution, Iodine swab
- ✓ Several different types of food for testing (such as a hard-boiled egg, bread, potato, pasta, yogurt, cookies or cheese)
- ✓ Eyedropper
- ✓ Small plastic cups
- ✓ Safety glasses

> **CAUTION:**
> The chemicals used in this experiment will stain clothing. Please use protective gear or wear old clothing when performing this experiment.

Procedure:
1. Read the introduction to this experiment and write down which of the test foods you think contain starch and which ones you think contain sugar.
2. Test each of your sample foods using the two tests below to see if they contain starch or sugar. (***Note***—*Directions for making the sample solution for the sugar test are on the experiment sheet on SG pg. 258.*)
   - ☑ **Starch test** – Cut off a sample of the food, then use an iodine swab to rub some solution onto it. If the iodine changes color to blue, then starch is present. If it remains the same brown color, no starch is present. Write down your results in the chart on your experiment sheet. Then, repeat this procedure for each of the test samples.
   - ☑ **Sugar Test** – Add 40 drops of the sample solution to a small plastic cup and put on your safety glasses. Add 10 drops of Benedict's solution and mix well. If the solution turns green, yellow or brick-red, sugar is present in some concentration. If the solution remains blue, there is no sugar in the sample solution.
3. Draw conclusions and complete the experiment sheet.

Vocabulary & Memory Work
- ☐ Vocabulary: carbohydrates, lipids, minerals, protein, starch, vitamins
- ☐ Memory Work—This week, add the following elements to what you are working on memorizing:
  - ✓ 101-Md-Mendelevium, 102-No-Nobelium, 103-Lr-Lawrencium, 104-Rf-Rutherfordium

Sketch: Food Chart
- Write down the food you eat in a day and research whether or not the item is a protein, a carbohydrate or a lipid. Then use that information to create a chart or poster that depicts what you have discovered.

Writing
- Reading Assignment: *DK Encyclopedia of Science* pp. 78-79 (Chemistry of Food)
- Additional Research Readings:
  - Chemistry of Food: *KSE* pp. 188-189, Food Industry: *DK EOS* pp. 92-93, Food: *UIDS* pp. 204-205

Dates
- 1813 – The first book on food chemistry, entitled *Elements of Agricultural Chemistry*, is published by Sir Humphrey Davy.

Sketch Lesson 3

Experiment: Is it a sugar or a starch?

Introduction

There is a whole branch of chemistry dedicated to the science of food. Scientists in this field study the chemical processes that are associated with food production and preservation. Some of the most basic tests that these chemists use are to determine whether or not starch or sugar is present in a given sample of food. In this week's experiment, you are going to perform these tests on several food samples from your own kitchen.

Hypothesis

___

___

___

Materials

___	___
___	___
___	___
___	___
___	___

How to make the sample solutions for the sugar test

1. Mix 4 to 5 tablespoons of the sample with about a cup of warm water and stir well.
2. Then, let the solution sit for 5 minutes or so before mixing it again. If there are solids, be sure to crush them up into small bits and let them settle before using the sample solution.

Procedure

___

___

___

___

___

___

Observations and Results

Sample	**Starch Test** *(positive or negative)*	**Sugar Test** *(positive or negative)*

Conclusion

Written Assignment Lesson 3

Discussion Questions

1. What are the groups of chemicals that are essential to life and why do we need them?
2. What do the chemicals in food do?
3. What happens to a protein when it is cooked?
4. How do we obtain the minerals our body needs?
5. How do we get the vitamins our body needs?
6. What happens to a starch molecule when it is cooked?

Written Assignment Lesson 3

Student Assignment Sheet Lesson 4
Fermentation

Experiment: Does temperature affect fermentation?

> **CAUTION**
> **DO NOT DO THIS EXPERIMENT WITHOUT ADULT SUPERVISION!**
> Boiling hot water can cause severe damage. Be sure to use the proper safety gear.

Materials:
- ✓ Yeast
- ✓ Water
- ✓ Sugar
- ✓ 3 bottles
- ✓ 3 balloons
- ✓ Instant read thermometer
- ✓ Pot
- ✓ Hot mitt

Procedure:
1. Read the introduction to this experiment and make a hypothesis.
2. Add 1 TBSP (8.5 g) of yeast and 1 TBSP (15 g) of sugar into each of the bottles. Label them #1 through #3.
3. Fill a pot with 4 cups of water and attach the instant read thermometer to the side. Slowly heat the water. Once it reaches 75° F (24° C), use the hot mitt to pour out one cup (240 mL) of water and then replace the pot on the burner. Quickly add the cup of 75° F (24° C) water to bottle #1, swirl the contents until mixed and place the balloon over the top.
4. Once the water reaches 95° F (35° C), use the hot mitt to pour out one cup (240 mL) of water and then replace the pot on the burner. Quickly add the cup of 95° F (35° C) water to bottle #2, swirl the contents until mixed and place the balloon over the top.
5. Once the water reaches 135° F (57° C), use the hot mitt to pour out one cup (240 mL) of water and then replace the pot on the burner. Quickly add the cup of 135° F (57° C) water to bottle #3, swirl the contents until mixed and place the balloon over the top.
6. After 15 minutes, measure the circumference of each of the balloons and record the measurements.
7. Draw conclusions and complete the experiment sheet.

Vocabulary & Memory Work
- ☐ Vocabulary: alcohol, fermentation, yeast
- ☐ Memory Work—This week, add the following elements to what you are working on memorizing:
  - ✓ 105-Db-Dubnium, 106-Sg-Seaborgium, 107-Bh-Bohrium, 108-Hs-Hassium

Sketch: Fermentation
- Label the following: start, yeast, sugar and water are mixed, after 15 minutes, yeast convert the sugars into alcohol which is left at the bottom of the container, carbon dioxide gas forms

Writing
- Reading Assignment: *DK Encyclopedia of Science* pg. 80 (Fermentation)
- Additional Research Readings:
  - Alcohols: *UIDS* pp. 196-197
  - Food Industry: *DK EOS* pp. 92-93

Dates
- 3000 BC – Yeast is used to make alcoholic drinks, such as beer or wine through fermentation.
- 200 – Yogurt is made through the use of bacteria.

Sketch Lesson 4

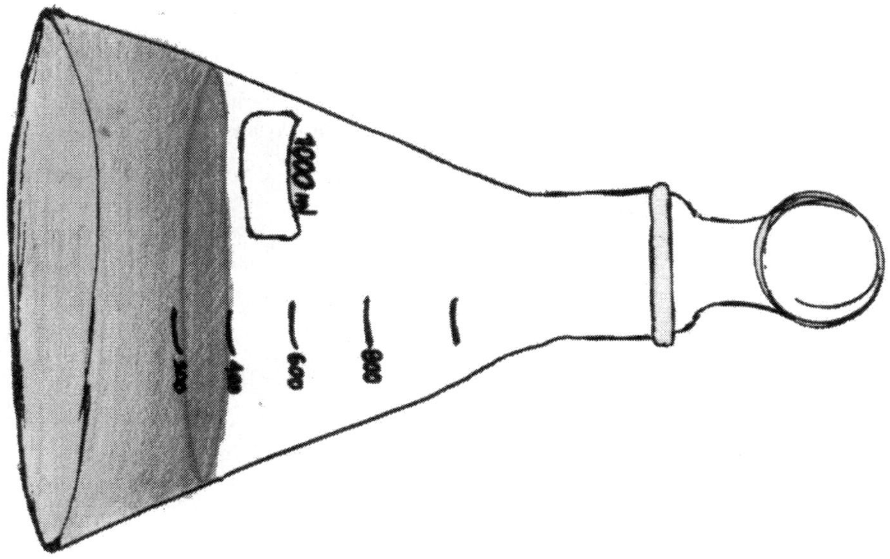

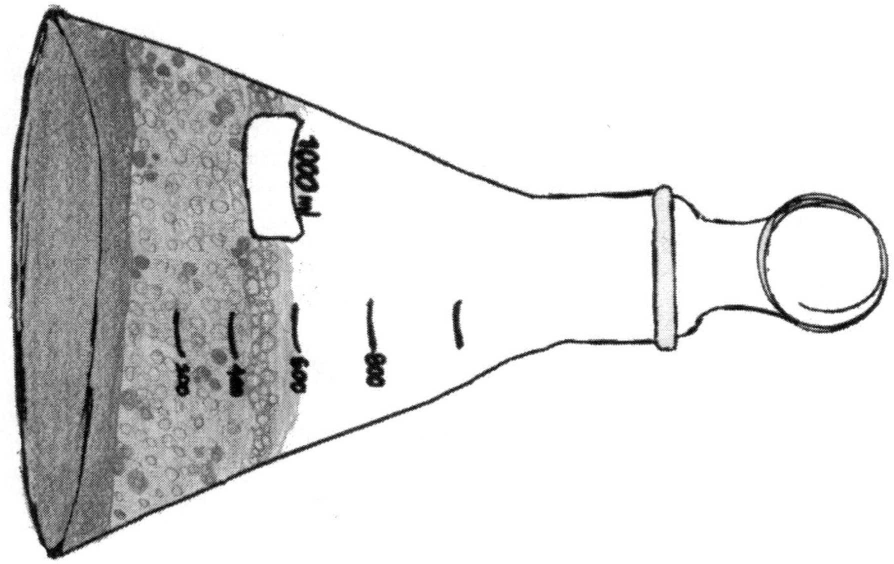

Experiment: Does temperature affect fermentation?

Introduction

Fermentation is the process in which the sugar in fruits or grains is converted by microbes into alcohol and carbon dioxide. Some of the most common fermentations are done by a group of microbes known as yeasts. Yeasts act as catalysts in the conversion of the glucose present in the fruit or grain to alcohol and carbon dioxide, which is shown in the equation below.

$C_6H_{12}O_6$ (glucose) $\longrightarrow$ $2C_2H_5OH$ (ethyl alcohol) + $2CO_2$ (carbon dioxide)

These fermentation processes require the right conditions to maximize their yield. In this week's experiment, you will test to see if temperature affects the amount of carbon dioxide the yeast can produce.

Hypothesis

____

____

____

Materials

______________________ ______________________

______________________ ______________________

______________________ ______________________

______________________ ______________________

______________________ ______________________

Procedure

____

____

____

____

____

____

____

Observations and Results

Temperature	Diameter of Balloon *(after 15 minutes)*
75°F (24°C)	
95°F (35°C)	
135°F (57°C)	

Conclusion

Written Assignment Lesson 4

Discussion Questions

1. What are some of the substances that fermentation is used to make?
2. How does fermentation with yeast change in the presence of oxygen?
3. How does fermentation in yogurt differ from the fermentation in blue cheese or from the fermentation that occurs when you make bread?

Written Assignment Lesson 4

Chemistry Unit 9

Chemistry of Industry

Unit 9: Chemistry of Industry
Vocabulary Sheet

Define the following terms as they are assigned on your Student Assignment Sheet.

1. Detergent – ____

____

____

2. Soap – ____

____

____

3. Adhesive – ____

____

____

4. Natural Polymer – ____

____

____

5. Synthetic Polymer – ____

____

____

6. Alloy – ____

____

____

7. Slag – ____

____

____

8. Hydrocarbon – ____

____

____

9. Raw Materials – ____

____

10. Dye – ______________________________

11. Emulsion – ______________________________

12. Pigment – ______________________________

13. Natural Fiber – ______________________________

14. Synthetic Fiber – ______________________________

Student Assignment Sheet Lesson 1
Soaps and Detergents

Experiment: Will soap and detergent always act the same?

Materials:

- ✓ Powdered detergent
- ✓ Liquid soap
- ✓ 2 large cups
- ✓ 2 small cups
- ✓ 2 bowls
- ✓ pH paper
- ✓ Vegetable oil
- ✓ Dirt
- ✓ Ketchup
- ✓ Plaster of Paris
- ✓ Water
- ✓ Straw
- ✓ Old t-shirt fabric

Procedure:

1. Read the introduction to this experiment and make a hypothesis.
2. Begin by making your solutions. First, mix 1 cup (240 mL) of water with 1 ½ tsp (5 g) of powdered detergent in a cup. Label the cup with "detergent solution". Then, mix 1 cup (240 mL) of water with 1 ½ tsp (7.5 mL) of liquid soap. Label the second cup with "soap solution".
3. Pour ½ cup (120 mL) of each solution into two different shallow bowls. Then, use the pH paper to test and record the pH of each of your solutions. Next, rub three pieces of t-shirt fabric with the dirt and ketchup. Set one of the pieces aside and place the other two in each of the bowls. Set aside to soak for 30 minutes.
4. Now add ¼ cup (60 mL) of each of the solutions to the small cups. Then, add 1 tsp (5 mL) of vegetable oil to each and agitate the solution. Observe and record what happened. Pour the mixtures out and rinse the cups well.
5. Add ¼ cup (60 mL) of each of the solutions to the small cups. Then, add 1 TBSP (10 g) of Plaster of Paris to each and agitate the solution. Using a straw blow into each of the solutions for 10 seconds, observe and record what happened. Let the solutions sit for 30 minutes and then observe what has happened.
6. After 30 minutes, remove the t-shirt fabric from each of the bowls, agitate by rubbing for 1-2 minutes and rinse well. Compare the two washed pieces with the unwashed one, observe and record your results.
7. Draw conclusions and complete the experiment sheet.

Vocabulary & Memory Work

- ☐ Vocabulary: detergent, soap
- ☐ Memory Work—Work on memorizing the following elements:
  - ✓ 109-Mt-Meitnerium, 110-Ds-Darmstadium, 111-Rg-Roentgenium, 112-Cn-Copernicium

Sketch: Soap and Detergent Molecules

- Label the following: soap molecule, carbon atom, oxygen atom, detergent molecule, sulfur atom, sodium atom

Writing

- Reading Assignment: *DK Encyclopedia of Science* pg. 82 (Chemical Industry) and pg. 95 (Soaps and Detergents)
- Additional Research Readings:
  - Soaps and Detergents: *KSE* pg. 187, Detergents: *UIDS* pp. 202-203, The Chemical Industry: *DK Chem* pp. 62-64

Sketch Lesson 1

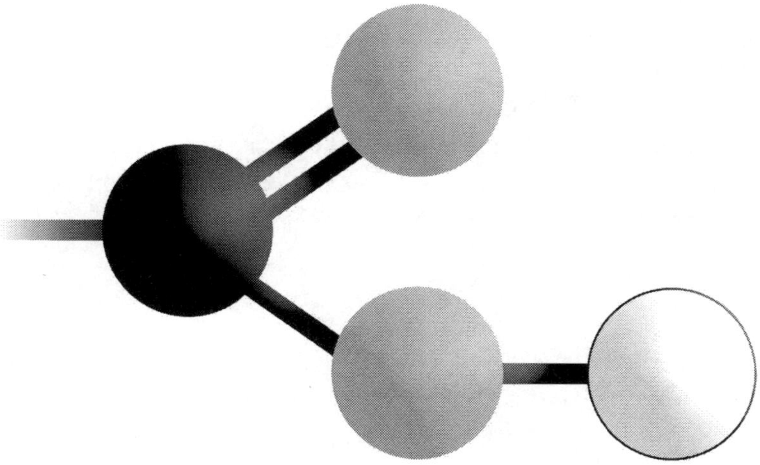

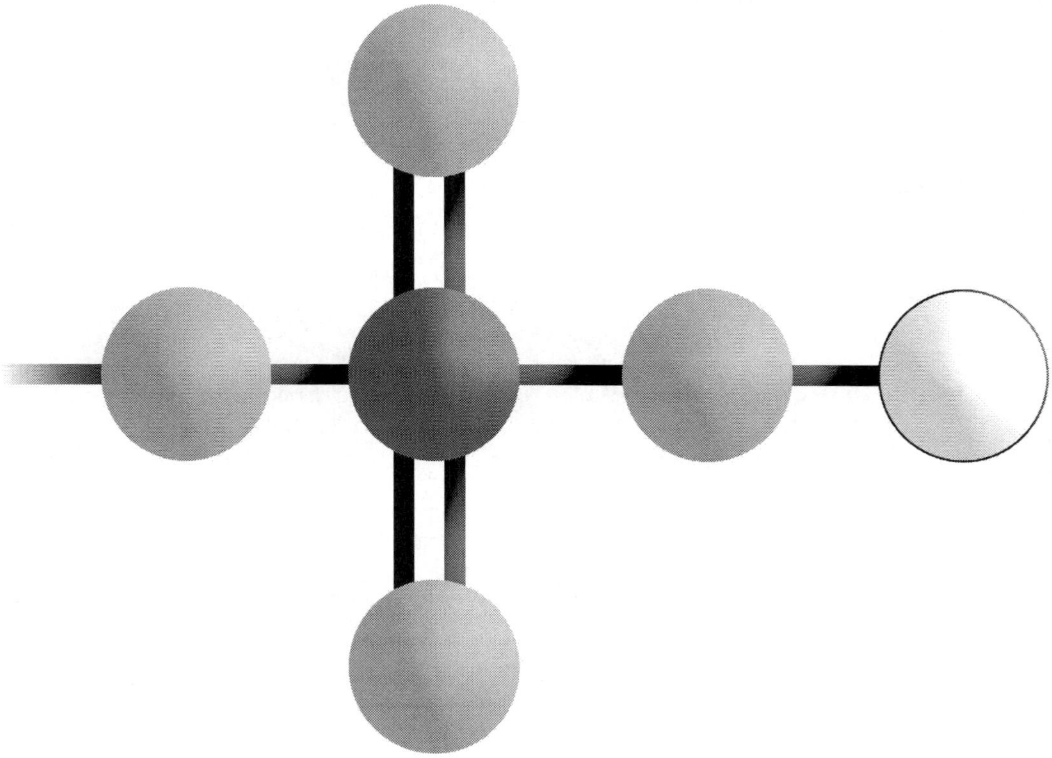

Experiment: Will soap and detergent always act the same?

Introduction

Detergents and soaps are a class of molecules that assist water in removing grease and oil from dirty material. Detergents are synthetic or organic substances that are not prepared from fats, while soaps are metallic salts derived from a fats. These molecules have a water-loving, polar head and a grease-loving, non-polar tail. So, they are able to dissolve in water, but at the same time they can attract grease and remove it from a surface. In this experiment, you are going to perform several tests to see how detergents and soaps work.

Hypothesis

____

____

Materials

______________________ ______________________

______________________ ______________________

______________________ ______________________

______________________ ______________________

______________________ ______________________

Procedure

____

____

____

____

____

____

____

____

____

____

Observations and Results

	pH	Mixing with Oil	Hard Water Test (Plaster of Paris)	Cleanliness Test
Soap Solution				
Detergent Solution				

Conclusion

Written Assignment Lesson 1

Discussion Questions

pg. 82, Chemical Industry

1. What happens to materials in the chemical industry?
2. What does every industrial plant need?

pg. 95, Soaps and Detergents

1. How do soaps and detergents differ?
2. How do soaps and detergents help to clean a surface?

Written Assignment Lesson 1

Student Assignment Sheet Lesson 2
Polymers and Adhesives

Science Fair Project

This week, you will complete step one and begin step two of your Science Fair Project. You will be choosing your topic, formulating a question and doing some research about that topic.

1. **Choose your topic** – You should choose a topic in the field of chemistry that interests you, such as electrolysis. Next, come up with several questions you have relating to that topic, (e.g. "Does the salt concentration affects electrolysis?" or "Does the strength of the battery affect the speed of electrolysis?"). Then, choose the one question you would like to answer and refine it (e.g. "Does salt affect the speed of hydrogen production in electrolysis?").
2. **Do Some Research** – Now that you have a topic and a question for your project, it is time to learn more about your topic so that you can make an educated guess (hypothesis) on the answer to your question. For the question stated above, you would need to research electrolysis and how hydrogen gas is produced during the process. Begin by looking up the topic in the references you have at home. Then, make a trip to the library to search for more on the topic. As you do your research, write any relevant facts you have learned on index cards and be sure to record the sources you use.

Vocabulary & Memory Work

- ☐ Vocabulary: adhesive, natural polymer, synthetic polymer
- ☐ Memory Work—Review the elements of the periodic table that you have memorized this year.

Sketch: Making a Polymer by Condensation

- Label the Following: the two monomers approach each other, a reaction occurs and several hydrogen and oxygen atoms split off, the two monomer chains join to form a polymer, the released hydrogen and oxygen atoms combine to form water

Writing

- Reading Assignment: *DK Encyclopedia of Science* pp. 100-101 (Polymers), pg. 106 (Adhesives)
- Additional Research Readings:
  - Polymers and Plastics: *KSE* pp. 205-206, *UIDS* pp. 200-201
  - The First Plastics: *DK Chem* pp. 54-55
  - Adhesives: *KSE* pg. 218

Dates to Enter

- 1850's – Alexander Parkes, an English chemist, synthesizes the first polymer.
- 1965 – Stephanie Kwolek, an American chemist for DuPont, creates the first Kevlar fibers, a polymer that is used to protect law enforcement officers from bullets.

Sketch Lesson 2

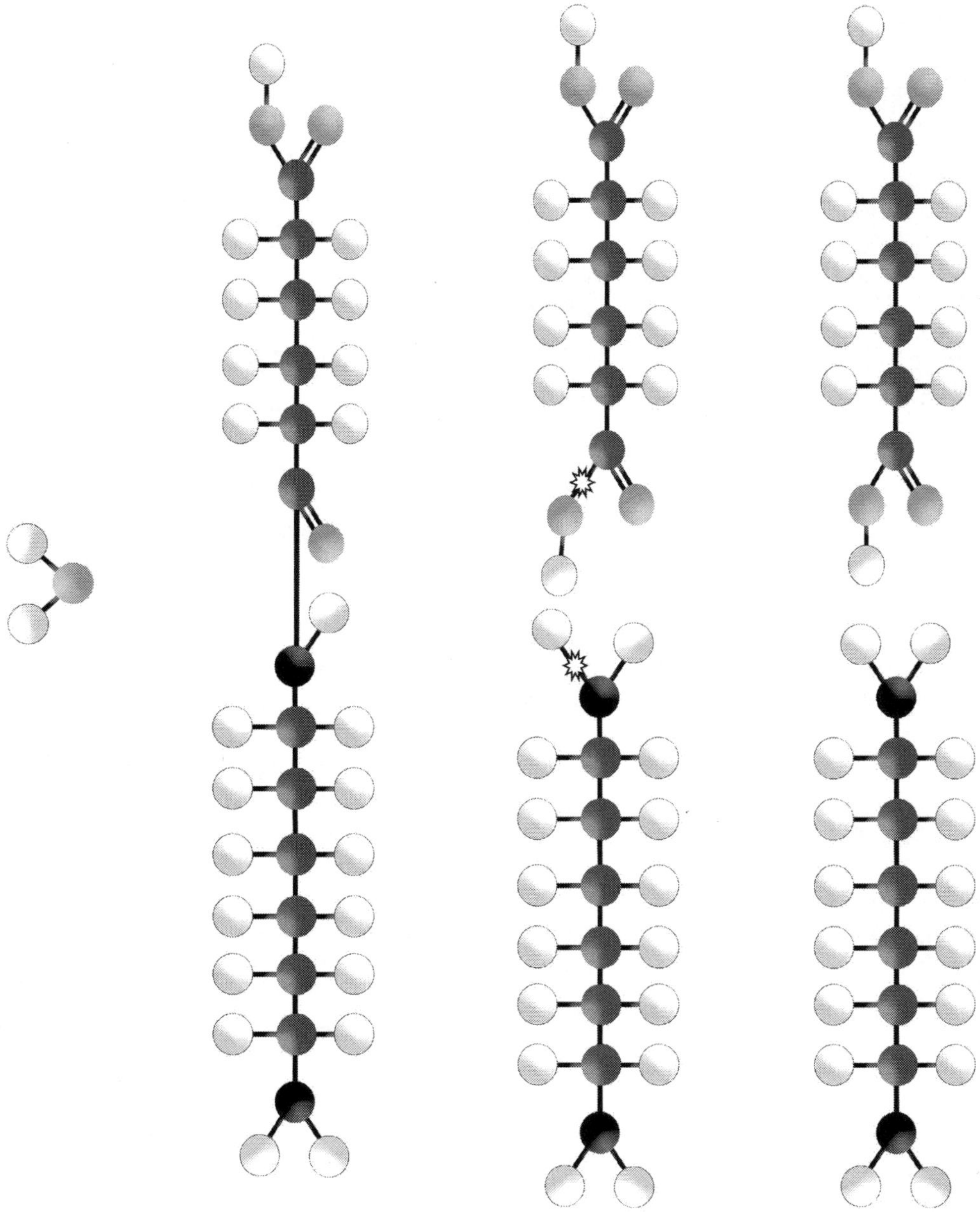

Science Fair Project Step 1: Choose a Topic

Key 1 ~ Decide on an area of chemistry.

What areas of chemistry are you interested in learning about?

____

____

____

____

Rank your interest in the different areas you listed and then circle the one area that you would like to use for your topic.

Key 2 ~ Develop several questions about the area of chemistry.

What questions would you like to answer about your area of chemistry? (***Note****—Remember that good questions begin with how, what, when, who, which, why, or where.*)

____

____

____

____

____

____

Key 3 ~ Choose a question to be the topic.

Write down the question that you will be using for your project.

____

____

____

Science Fair Project Step 2: Do Some Research

Key 1 ~ Brainstorm for research categories.

What categories are you going to research for your project?

1. ______________________________________

2. ______________________________

3. ______________________________

4. ______________________________

5. ______________________________

Key 2 ~ Research the categories.

Use the following template for your research cards:

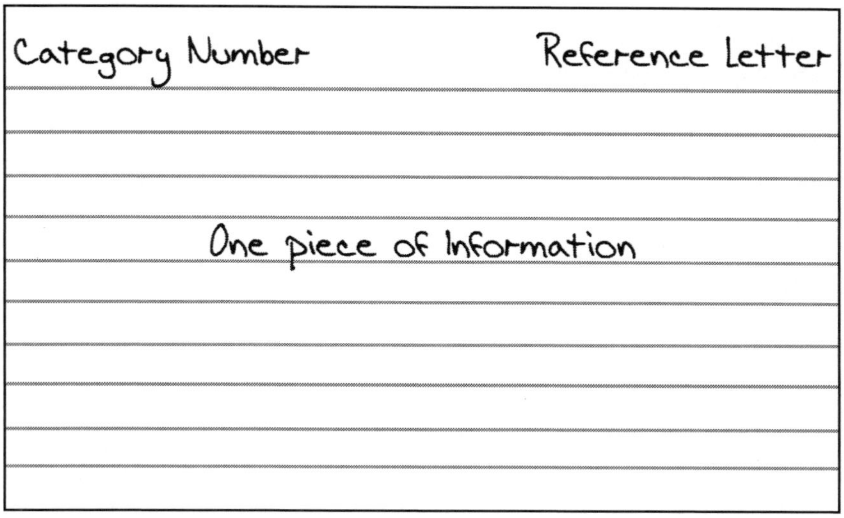

Record your sources below.

A. ______________________________

____

B. ______________________________

____

C. ______________________________

____

D. ______________________________

____

E. ______________________________

____

F. ______________________________

____

Written Assignment Lesson 2

Discussion Questions

pp. 100-101, Polymers

1. What are several examples of natural polymers? Synthetic polymers?
2. What is addition polymerization?
3. How are polymers made with condensation?
4. What are several ways that polymer pellets can be turned into useful materials?

pg. 106, Adhesives

1. How do adhesives work?
2. What is special about thermoplastic glues?
3. Why are some epoxy resins a two-part glue?

Written Assignment Lesson 2

Student Assignment Sheet Lesson 3
Iron, Steel, and Alloys

Science Fair Project

This week, you will complete steps two through four of your Science Fair Project. You will be finishing your research, formulating your hypothesis and designing your experiment.

2. **Do Some Research** – This week, you will finish your research. Then, organize your research index cards and write a brief report on what you have found out.
3. **Formulate a Hypothesis** – A hypothesis is an educated guess. For this step, you need to review your research and make an educated guess about the answer to your question. A hypothesis for the question asked in step one would be, "If the concentration of the salt is increased, then the rate that hydrogen gas is produced during electrolysis with increase."
4. **Design an Experiment** – Your experiment will test the answer to your question. It needs to have a control and several test groups. Your control will have nothing changed, while your test groups will change only one factor at a time. An experiment to test the hypothesis given above would be to make three different solutions for electrolysis with varying salt concentration for your test groups and use plain distilled water for the control group. Then, allow the electrolysis to run for each group until you collect a set amount of hydrogen gas. Each time, you would record the amount of time it took. If time allows, you can go ahead and begin your experiment this week.

Vocabulary & Memory Work

- ☐ Vocabulary: alloy, slag
- ☐ Memory Work—Review the elements of the periodic table that you have memorized this year.

Sketch: Blast Furnace Reaction

- Label the Following: limestone, coke, iron ore, heat, molten iron, carbon grabs the oxygen from the iron oxide to form carbon dioxide gas, limestone traps the impurities left in the iron ore to form slag

Writing

- Reading Assignment: *DK Encyclopedia of Science* pp. 84-85 (Iron and Steel), pg. 88 (Alloys)
- Additional Research Readings:
  - Iron: *KSE* pg. 198, *UIDS* pg. 174
  - Alloys: *KSE* pp. 202-203
  - Copper: *DK EOS* pg. 86, *KSE* pg. 199

Dates to Enter

- c. 3000 BC – The first alloy is created during the Bronze Age.
- 1856 – Henry Bessemer invents a way of removing most of the carbon from iron, creating steel.

Sketch Lesson 3

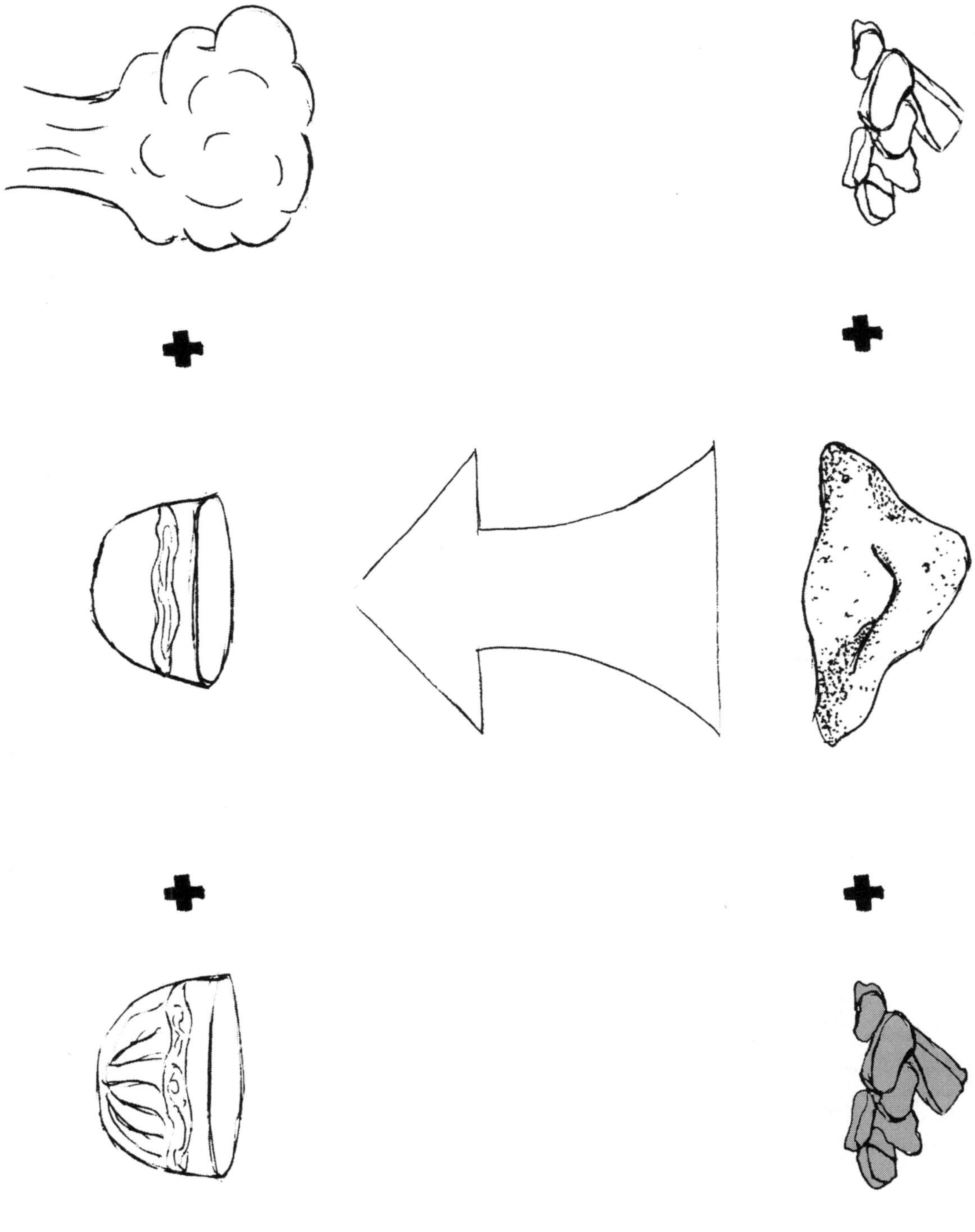

Science Fair Project Step 2: Do Some Research

Key 3 ~ Organize the information.

Organize the information for your report.

Key 4 ~ Write a brief report.

Write down what the order of your categories will be for your report.

1. ______________________________
2. ______________________________
3. ______________________________
4. ______________________________
5. ______________________________

On a separate sheet of paper write out a rough draft of your research report.

Science Fair Project Step 3: Formulate a Hypothesis

Key 1 ~ Review the Research.

Read over your research.

Key 2 ~ Formulate an Answer.

Write down your hypothesis for your science fair project.

____

____

____

____

____

Science Fair Project Step 4: Design an Experiment

Key 1 ~ Choose a Test.

What are some ways that you can test your hypothesis?

____

____

____

Key 2 ~ Determine the Variables.

What factor are we trying to test? (independent variable)

What factor will we use to measure the progress of our test? (dependent variable)

What factors do we need to keep constant so that they will not affect our results? (controlled variables)

Key 3 ~ Plan the Experiment.

What will the groups in your experiment be?

Control Group: ______

Test Group 1: ______

Test Group 2: ______

Test Group 3: ______

Test Group 4: ______

Write down the plan for your experiment.

Written Assignment Lesson 3

Discussion Questions

pp. 84-85, Iron and Steel

1. What is smelting?
2. What happens when iron ore is smelted in a blast furnace?
3. What is steel and why is it beneficial?

pg. 88, Alloys

1. Why are alloys better than a pure metal on its own?

Written Assignment Lesson 3

Student Assignment Sheet Lesson 4
Coal and Oil Products

Science Fair Project

This week, you will complete steps five and six of your Science Fair Project. You will carry out the experiment and record your observations and results.

5. **Perform the Experiment** – This week, you will perform the experiment you designed last week. Be sure to take pictures along the way as well as record your observations and results. (***Note***—*Observations are a record of the things you see happening in your experiment. For instance, an observation would be that the hydrogen gas from the control group solution filled almost half of the test tube. Results are specific and measurable. For instance, results could be that you collected 10 cm of hydrogen gas in 20 minutes from the electrolysis test #1. Observations are generally recorded in journal form, while results can be compiled into tables, charts and graphs or relayed in paragraph form.*)
6. **Analyze the Data** – Once you have compiled your observations and results, you can use them to answer your question. You need to look for trends in your data and make conclusions from that. A possible conclusion to the electrolysis experiment would be, "The more salt present in the electrolyte solution, the quicker the hydrogen gas is produced in electrolysis." If your hypothesis does not match your conclusion or your were not able to answer your question using the results from your experiment, you may need to go back and do some additional experimentation.

Vocabulary & Memory Work

- Vocabulary: hydrocarbons
- Memory Work—Review the elements of the periodic table that you have memorized this year.

Sketch: Fractional Distillation of Crude Oil

- Label the Following: crude oil is heated, residue falls to the bottom, gas oil condenses, kerosene condenses, naphtha condenses, gasoline condenses, refinery gases remain

Writing

- Reading Assignment: *DK Encyclopedia of Science* pg. 96 (Coal Products), pp. 98-99 (Oil Products)
- Additional Research Readings:
    - Oil and Refining: *KSE* pp. 210-211
    - Coal: *KSE* pg. 213
    - Petroleum: *UIDS* pp. 198-199

Dates to Enter:

- 347 – Oil wells are first drilled in China. The oil was naturally refined and used as a building adhesive, for water-proofing and as a fuel.
- 1850's – The first commercial oil refinery appears.

Sketch Lesson 4

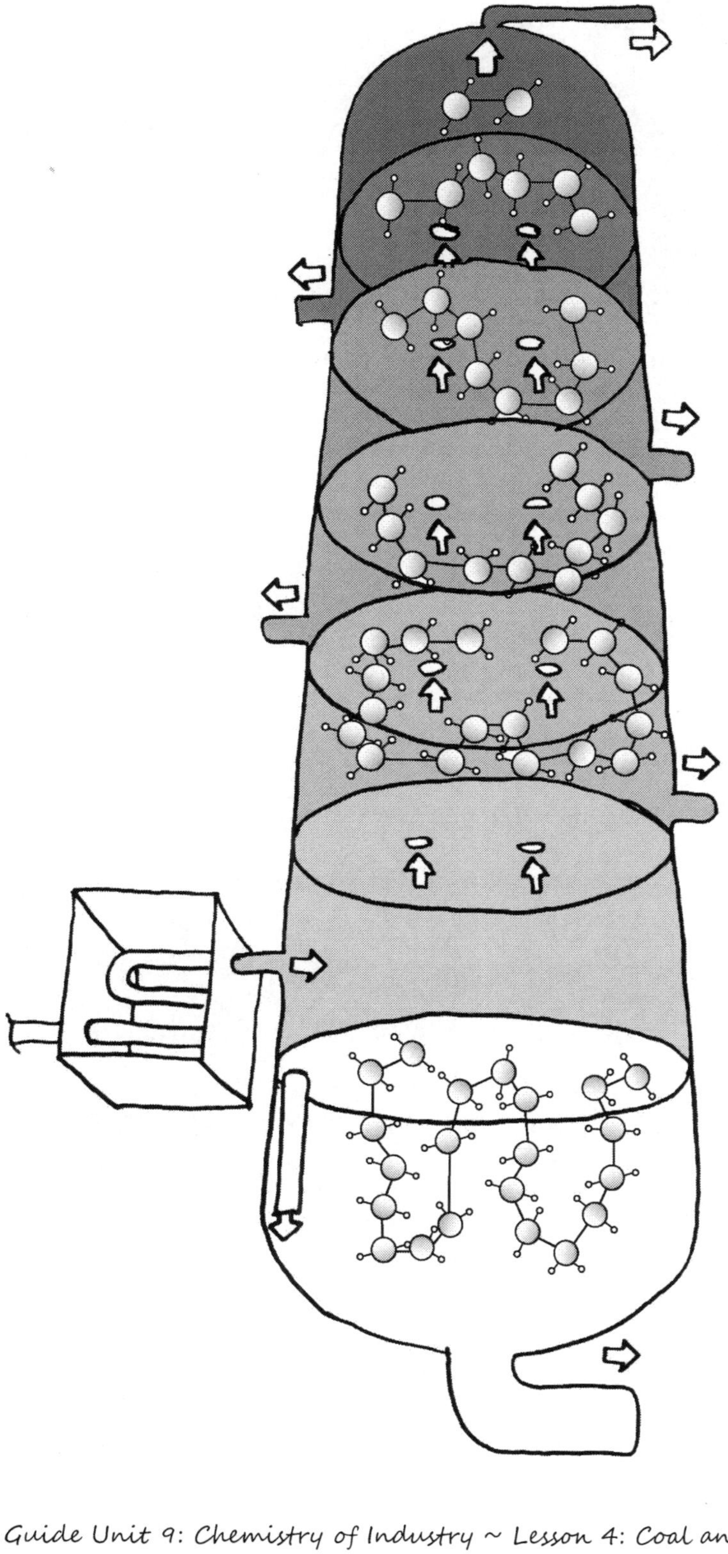

Science Fair Project Step 5: Perform the Experiment

Key 1 ~ Get ready for the experiment.

When do you plan to run your experiment?

From ______________________________ to ______________________________.

Purchase and gather your materials, and prepare any of the materials that need to be pre-made.

Key 2 ~ Run the experiment.

What things do you need to remember to do each day?

Take pictures of the experiment every day or for every trial.

____

____

____

____

____

____

____

____

____

Key 3 ~ Record any Observations and Results.

Record your observations and results on a separate sheet of paper.

Science Fair Project Step 6: Analyze the Data

Key 1 ~ Review and organize the data.

What trends did you recognize in your observations?

____

____

____

What information did you interpret from your results?

Key 2 ~ State the answer.

After reviewing your data, write the answer to your question. (***Note****—Your statement should begin with "I found that…" or "I discovered that…"*)

Key 3 ~ Draw several conclusions.

Answer the following questions:

- ✓ Was my hypothesis proven true? (***Note****—If your hypothesis was proven false, be sure to state why you think it was proven false.*)
- ✓ Did you have any problems or difficulties when performing your experiment?
- ✓ Did anything interesting happen that you would like to share?
- ✓ Can you think of any other things related to your project that you would like to test in the future?

Now take your answer from key two and your answers from key three to write your conclusion on a separate sheet of paper. Your paragraph should be four to six sentences in length.

Written Assignment Lesson 4

Discussion Questions

pg. 96, Coal Products

1. How is coke produced and what are the side products?
2. Other than a source of energy, what can coal products be used for?

pp. 98-99, Oil Products

1. What does crude oil consist of?
2. How does fractional distillation work?
3. What does catalytic cracking do?

Written Assignment Lesson 4

Student Assignment Sheet Lesson 5
Chemistry in Farming

Science Fair Project

This week, you will complete steps seven and eight of your Science Fair Project. You will be writing and preparing a presentation of your Science Fair Project.

7. **Create a Board** – This week, you will be creating a visual representation of your science fair project that will serve as the centerpiece of your presentation. You will begin by planning the look of your board, then move onto preparing the information and finally you will pull it all together.
8. **Give a Presentation** – After you have completed your presentation board, determine if you would like to include part of your experiment in your presentation. Then, prepare a 5 minute talk about your project, be sure to include the question you tried to answer, your hypothesis, a brief explanation of your experiment and the results plus the conclusion to your project. Be sure to arrive on time for your presentation. Set up your project board and any other additional materials. Give your talk and then ask if there are any questions. Answer the questions and end your time by thanking whoever has come to listen to your presentation.

Vocabulary & Memory Work

- Vocabulary: raw materials
- Memory Work—Review the elements of the periodic table that you have memorized this year.

Sketch: Ammonia Production

- Label the Following: pure hydrogen, pure nitrogen, compressor, catalyst chamber, cooling chamber, gas is squeezed and heated, the hot gases touch the iron pellets and are drawn together, the gases are cooled until liquid ammonia pours off, the unreacted gases are recycled

Writing

- Reading Assignment: *DK Encyclopedia of Science* pg. 90 (Ammonia), pg. 91 (Chemistry in Farming)
- Additional Research Readings:
  - Food Production: *KSE* pp. 230-231

Dates

- 1908 – Fritz Haber designs the first apparatus to produce ammonia. Later, his design is scaled up for industrial purposes by Carl Bosch.

Sketch Lesson 5

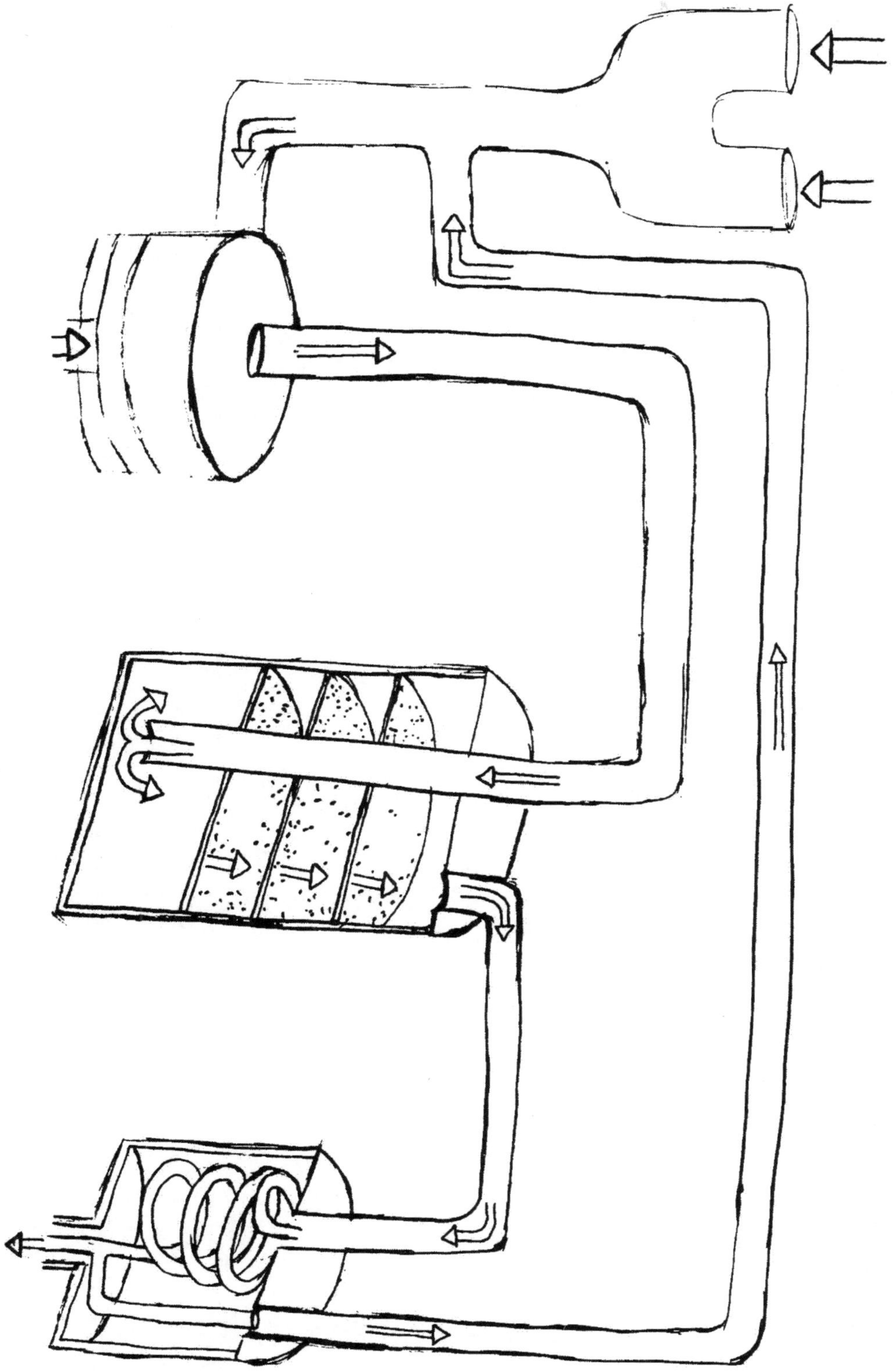

Science Fair Project Step 7: Create a Board

Key 1 ~ Plan out the board.

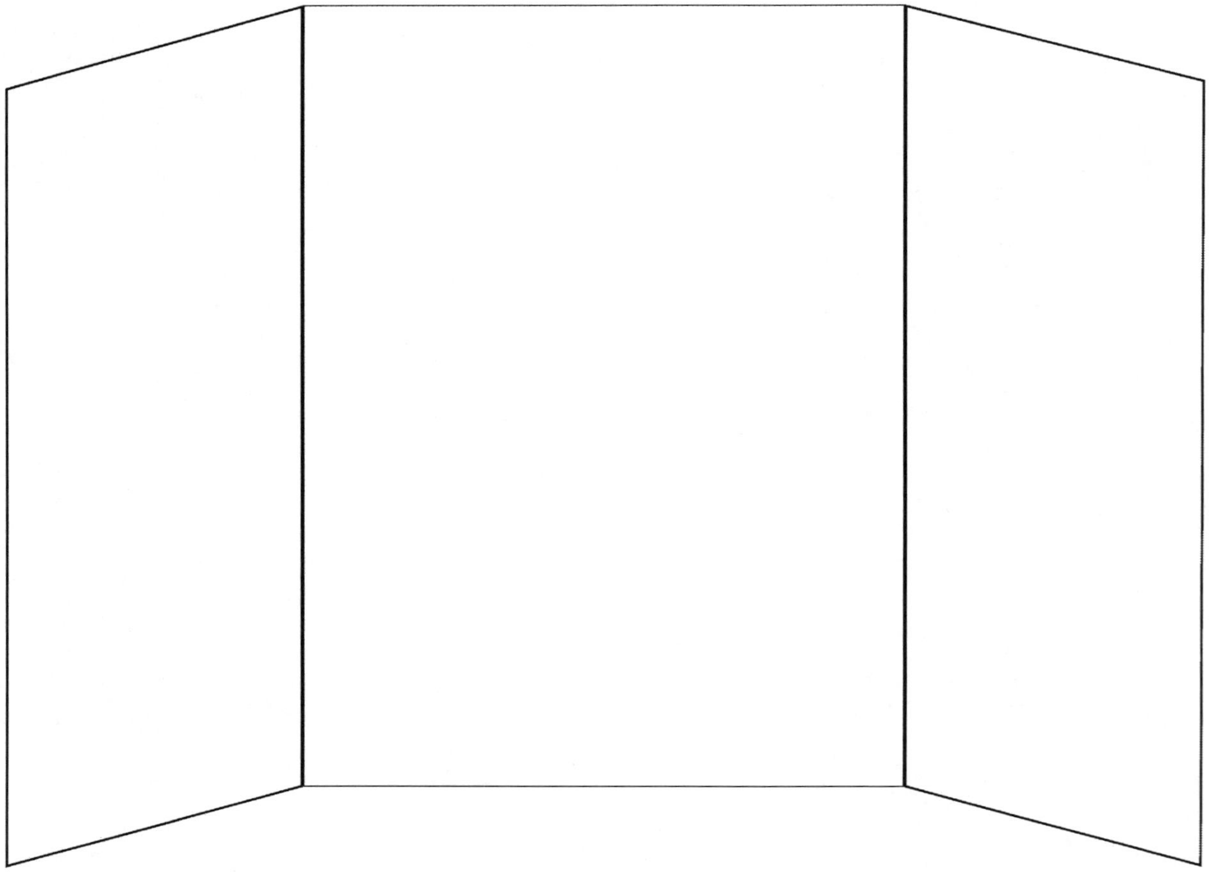

Key 2 ~ Prepare the information.

Type up the following information for your presentation board:

- ☐ Introduction
- ☐ Hypothesis
- ☐ Research
- ☐ Materials
- ☐ Procedure
- ☐ Results
- ☐ Conclusion

Key 3 ~ Put the board together.

☐ Put the decorative elements on your project board.

☐ Print out and attach your information paragraphs.

☐ Add the title to your science fair project board.

Science Fair Project Step 8: Give a Presentation

Key 1 ~ Prepare the presentation.

Write down the outline for your presentation on a separate sheet of paper.

Key 2 ~ Practice the presentation.

☐ Practice your presentation in front of a mirror several times.

☐ Practice your presentation with your teacher.

Key 3 ~ Share the presentation.

Keep the following tips in mind for your presentation:

✓ Arrive on time for your presentation.
✓ Set up your project board and any other additional materials.
✓ Give your talk and then ask if there are any questions.
✓ Answer the questions and end your time by thanking whomever has come to listen to your presentation.

Written Assignment Lesson 5

Discussion Questions

pg. 90, Ammonia

1. What are the raw materials for producing ammonia?
2. What is the catalyst used in ammonia production?
3. Who is responsible for inventing the process by which ammonia is produced?

pg. 91, Chemistry of Farming

1. What does the chemical industry provide for farmers?
2. How does organic farming differ?

Written Assignment Lesson 5

Student Assignment Sheet Lesson 6
Dyes, Pigments, and Cosmetics

Scientist Biography Report Project

This week, you will complete step one and begin step two of your Scientist Biography Report Project. You will be choosing the scientist you would like to learn more about and begin your research. The instructions for this week's assignments are on the following Scientist Biography Report sheets.

Vocabulary & Memory Work

- Vocabulary: dye, emulsion, pigment
- Memory Work—Review the elements of the periodic table that you have memorized this year.

Sketch: Emulsion

- Label the following: water layer, oil layer, addition of an emulsifier, water molecule, oil molecule, emulsifying molecules, oil-in-water emulsion

Writing

- Reading Assignment: *DK Encyclopedia of Science* pg. 102 (Dyes and Pigments) and pg. 103 (Cosmetics)
- Additional Research Readings:
  - Pigments and Dyes: *KSE* pg. 219

Dates

- 5000 BC – The Egyptians begin to use cosmetics to alter their looks.
- 2600 BC – The earliest known written record of the use of dyes in China.
- 400 – Murex, the mollusk from which the Romans make a purple dye, becomes scarce due to huge demand and over harvesting.
- 1856 – William Henry Perkin discovers the first synthetic dye while searching for a cure for malaria.

Sketch Lesson 6

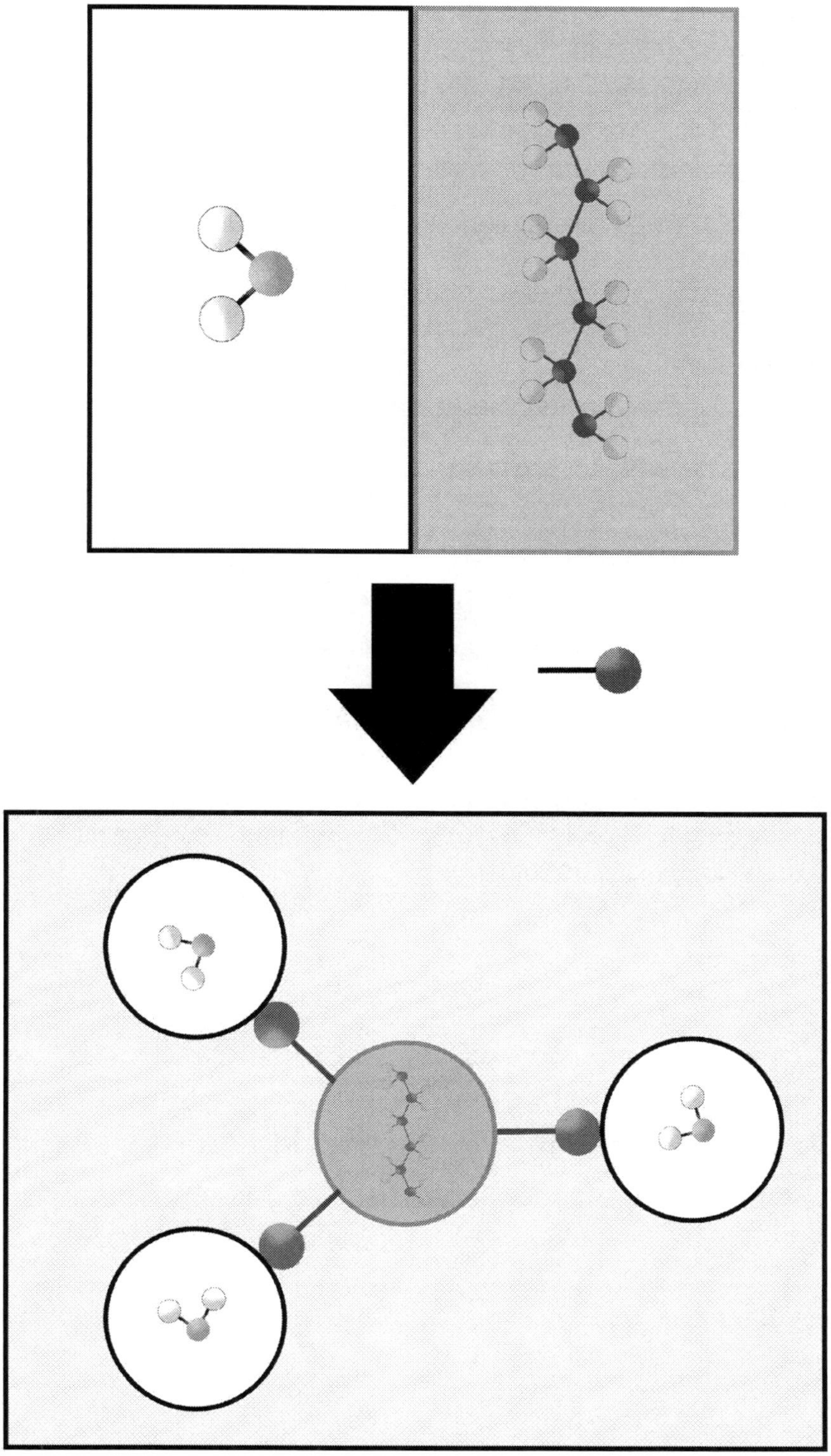

Scientist Biography Report

Step 1: Choose a Scientist

During the next three weeks, you are going to be researching and learning more about a scientist that has contributed to the field of chemistry. This week, you need to begin your scientist biography project by choosing which scientist you will research. You can choose one of the scientists mentioned in the "Dates" sections or you can choose one that has interested you.

The scientist I am going to study is:

___

Step 2: Research the Scientist

Once you have chosen the scientist you would like to study, you can begin your research. Begin by looking for a biography on your chosen scientist at the library. Then, look for articles on the chemist in magazines, newspapers, encyclopedias or on the Internet. You will need to know the following about your scientist to write your report:

- ☑ **Biographical information on the scientist (***i.e. where they were born, their parents, siblings, and how they grew up***);**
- ☑ **The scientist's education (***i.e. where they went to school, what kind of student they were, what they studied, and so on***);**
- ☑ **Their scientific contributions (***i.e. research that they participated in, any significant discoveries they made, and the state of the world at the time of their contributions***).**

As you read over the material you have gathered, be sure to write down any facts you glean in your own words. You can do this on the sheet below or on separate index cards. You will have two weeks to complete your research, so plan accordingly.

My Research

___

___

___

___

___

___

___

___

___

Written Assignment Lesson 6

Discussion Questions

pg. 102, Dyes and Pigments

1. What is the difference between dyes and pigments?
2. How do natural and synthetic dyes differ?
3. What happens when paint dries?

pg. 103, Cosmetics

1. What are modern day cosmetics made from?
2. Why are emulsions so important to the cosmetic industry?

Written Assignment Lesson 6

Student Assignment Sheet Lesson 7
Fibers and Paper

Scientist Biography Report Project

This week, you will complete step two and three of the Scientist Biography Report Project. You will be finishing up your research and organizing the information you have gathered into an outline. The instructions for this week's assignments are on the following Scientist Biography Report sheets.

Vocabulary & Memory Work

- ☐ Vocabulary: natural fiber, synthetic fiber
- ☐ Memory Work—Review the elements of the periodic table that you have memorized this year.

Sketch

- There is no sketch this week.

Writing

- Reading Assignment: *DK Encyclopedia of Science* pg. 107 (Fibers) and pg. 108 (Paper)
- Additional Research Readings:
  - Wood and Paper: *KSE* pg. 206
  - Fibers: *KSE* pg. 207

Dates

- c. 105 – The Chinese begin making paper from wood pulp.

Scientist Biography Report

Step 2: Research the Scientist

This week, you need to wrap-up the research you have done on your scientist. Read over your notes and make sure that you have at least five pieces of information for each of the categories below:

- ☑ **Biographical information on the scientist** (*i.e. where they were born, their parents, siblings, and how they grew up*)**;**
- ☑ **The scientist's education** (*i.e. where they went to school, what kind of student they were, and what they studied*)**;**
- ☑ **Their scientific contributions** (*i.e. research that they participated in, any significant discoveries they made, and the state of the world at the time of their contributions*)**.**

Also, make sure that the recorded list of the resources you have used is complete.

My Research

Step 3: Create an Outline

Now that your research is completed, you are ready to begin the process of writing a report on your chosen scientist. This week, you are going to organize the notes you took during step two into a formal outline which you will use next week to write the rough draft of your report. Use the outline template provided below as a guide. You should include information on why you chose the particular scientist in your introduction section. For the conclusion section of the outline, you need to include why you believe someone else should learn about your chosen scientist and your impression of the scientist (i.e. *Did you like the scientist? Do you feel that they made a significant impact on the field of chemistry?*).

Scientist Biography Outline

I. Introduction and Biological Information on the Scientist

A. ______________________________

____

B. ______________________________

____

C. ______________________________

____

D. ______________________________

____

E. ______________________________

____

F. ______________________________

____

II. The Scientist's Education

A. ______________________________

____

B. ______________________________

____

C. ______________________________

____

D. ______________________________

E.

F.

III. The Scientist's Contributions

A.

1.

2.

3.

4.

B.

1.

2.

3.

4.

C.

1. ______________________________

____

2. ______________________________

____

3. ______________________________

____

4. ______________________________

____

IV. Conclusion

A. ______________________________

____

B. ______________________________

____

C. ______________________________

____

D. ______________________________

____

E. ______________________________

____

F. ______________________________

____

Written Assignment Lesson 7

Discussion Questions

pg. 107, Fibers

1. What are several types of natural fibers? Synthetic fibers?
2. How is nylon made?

pg. 108, Paper

1. How is paper made?

Student Assignment Sheet Lesson 8
Ceramics and Glass

Scientist Biography Report Project

This week, you will complete steps four and five of the Scientist Biography Report Project. You will use your outline from last week to create a rough draft of your report. Then, you will edit and revise your paper to create the final draft. The instructions for this week's assignments are on the following Scientist Biography Report sheets.

Vocabulary & Memory Work

- Vocabulary: There is no vocabulary this week.
- Memory Work—Review the elements of the periodic table that you have memorized this year.

Sketch

- There is no sketch this week.

Writing

- Reading Assignment: *DK Encyclopedia of Science* pg. 109 (Ceramics) and pg. 110 (Glass)
- Additional Research Readings:
  - Glass: *KSE* pg. 208
  - Ceramics: *KSE* pg. 209

Dates

- 5000 BC – Egyptians use pottery (ceramics) and glazes.
- 3000 BC – The first glass is thought to be made during the Bronze Age.

Scientist Biography Report

Step 4: Write a Rough Draft

Last week, you created a formal outline for your scientist biography report; now, it is time to take that outline and turn it into a rough draft. Simply take the points on your outline and combine or add in sentence openers to create a cohesive paragraph. Here's what your rough draft should look like:

- ☑ Paragraph 1 (*from outline point I*): introduce the scientist;
- ☑ Paragraph 2 (*from outline point II*): tell about the scientist's education;
- ☑ Paragraph 3-5 (*from outline point III*): share the scientist's contributions (*one paragraph for each contribution*);
- ☑ Paragraph 6 (*from outline point IV*): share your thoughts on the scientist and why someone should learn about him or her.

You can choose to hand write or type up your rough draft on a separate sheet of paper. However, keep in mind that you will need a typed version for step five.

Step 5: Revise to Create a Final Draft

Now that you have a typed, double-spaced rough draft, look over it one more time to make any changes you would like. Then, have your teacher or one of your peers look over the paper for you to correct any errors and bring clarity to any of the unclear sections. Once this is complete, make the necessary changes to your paper to create your final draft. Print out your final paper and include it on the next page.

Scientist Biography Report: Final Draft

Scientist Biography Report: Final Draft

Written Assignment Lesson 8

Discussion Questions

pg. 109, Ceramics

1. What are the two groups of ceramics?
2. Why are ceramics good materials for engines?

pg. 110, Glass

1. What is glass made from?
2. Why is glass so useful?
3. How is float glass made?

Written Assignment Lesson 8

Appendix

Astronomy Memory Work

Unit 1

Types of Stars

1. Blue giant – A large, hot star off the main sequence.
2. Red giant – An older star with a cooler outer layer.
3. Neutron star – The tightly-packed collapsed core of a larger star.
4. Main-sequence star – A star plotted in the left-to-right band across the HR diagram.
5. Black hole – A gravitationally dense region of space-time where nothing can escape, not even light.
6. White dwarf – A stellar core remnant of a low to medium mass star.
7. Red dwarf – A cool, small star on the main sequence.

Constellations of the Zodiac

1. Aquarius
2. Aries
3. Cancer
4. Capricorn
5. Gemini
6. Leo
7. Libra
8. Pisces
9. Sagittarius
10. Scorpio
11. Taurus
12. Virgo

Unit 2

Planet Order *(along with the planet's gravity relative to Earth)*

1. Sun
2. Mercury (Gravity: 0.38)
3. Venus (Gravity: 0.90)
4. Earth (Gravity: 1 or 9.8 m/s2) [Moon (Gravity: 0.17)]
5. Mars (Gravity: 0.38)
6. Jupiter (Gravity: 2.34)
7. Saturn (Gravity: 0.93)
8. Uranus (Gravity: 0.90)
9. Neptune (Gravity: 1.13)

Chemistry Memory Work

The Elements of the Periodic Table

- ✓ 1-H-Hydrogen
- ✓ 2-He-Helium
- ✓ 3-Li-Lithium
- ✓ 4-Be-Beryllium
- ✓ 5-B-Boron
- ✓ 6-C-Carbon
- ✓ 7-N-Nitrogen
- ✓ 8-O-Oxygen
- ✓ 9-F-Fluorine
- ✓ 10-Ne-Neon
- ✓ 11-Na-Sodium
- ✓ 12-Mg-Magnesium
- ✓ 13-Al-Aluminum
- ✓ 14-Si-Silicon
- ✓ 15-P-Phosphorus
- ✓ 16-S-Sulfur
- ✓ 17-Cl-Chlorine
- ✓ 18-Ar-Argon
- ✓ 19-K-Potassium
- ✓ 20-Ca-Calcium
- ✓ 21-Sc-Scandium
- ✓ 22-Ti-Titanium
- ✓ 23-V-Vanadium
- ✓ 24-Cr-Chromium
- ✓ 25-Mn-Manganese
- ✓ 26-Fe-Iron
- ✓ 27-Co-Cobalt
- ✓ 28-Ni-Nickel
- ✓ 29-Cu-Copper
- ✓ 30-Zn-Zinc
- ✓ 31-Ga-Gallium
- ✓ 32-Ge-Germanium
- ✓ 33-As-Arsenic
- ✓ 34-Se-Selenium
- ✓ 35-Br-Bromine
- ✓ 36-Kr-Krypton
- ✓ 37-Rb-Rubidium
- ✓ 38-Sr-Strontium
- ✓ 39-Y-Yttrium
- ✓ 40-Zr-Zirconium
- ✓ 41-Nb-Niobium
- ✓ 42-Mo-Molybdenum
- ✓ 43-Tc-Technetium
- ✓ 44-Ru-Ruthenium
- ✓ 45-Rh-Rhodium
- ✓ 46-Pd-Palladium
- ✓ 47-Ag-Silver
- ✓ 48-Cd-Cadmium
- ✓ 49-In-Indium
- ✓ 50-Sn-Tin
- ✓ 51-Sb-Antimony
- ✓ 52-Te-Tellurium
- ✓ 53-I-Iodine
- ✓ 54-Xe-Xenon
- ✓ 55-Cs-Cesium
- ✓ 56-Ba-Barium
- ✓ 57-La-Lanthanum
- ✓ 58-Ce-Cerium
- ✓ 59-Pr-Praseodymium
- ✓ 60-Nd-Neodymium
- ✓ 61-Pm-Promethium
- ✓ 62-Sm-Samarium
- ✓ 63-Eu-Europium
- ✓ 64-Gd-Gadolinium
- ✓ 65-Tb-Terbium
- ✓ 66-Dy-Dysprosium
- ✓ 67-Ho-Holmium
- ✓ 68-Er-Erbium
- ✓ 69-Tm-Thulium
- ✓ 70-Yb-Ytterbium
- ✓ 71-Lu-Lutetium
- ✓ 72-Hf-Hafnium
- ✓ 73-Ta-Tantalum
- ✓ 74-W-Tungsten
- ✓ 75-Re-Rhenium
- ✓ 76-Os-Osmium
- ✓ 77-Ir-Iridium
- ✓ 78-Pt-Platinum
- ✓ 79-Au-Gold
- ✓ 80-Hg-Mercury

Chemistry Memory Work

The Elements of the Periodic Table

- ✓ 81-Tl-Thallium
- ✓ 82-Pb-Lead
- ✓ 83-Bi-Bismuth
- ✓ 84-Po-Polonium
- ✓ 85-At-Astatine
- ✓ 86-Rn-Radon
- ✓ 87-Fr-Francium
- ✓ 88-Ra-Radium
- ✓ 89-Ac-Actinium
- ✓ 90-Th-Thorium
- ✓ 91-Pa-Protactinium
- ✓ 92-U-Uranium
- ✓ 93-Np-Neptunium
- ✓ 94-Pu-Plutonium
- ✓ 95-Am-Americium
- ✓ 96-Cm-Curium
- ✓ 97-Bk-Berkelium
- ✓ 98-Cf-Californium
- ✓ 99-Es-Einsteinium
- ✓ 100-Fm-Fermium
- ✓ 101-Md-Mendelevium
- ✓ 102-No-Nobelium
- ✓ 103-Lr-Lawrencium
- ✓ 104-Rf-Ruterfordium
- ✓ 105-Db-Dubnium
- ✓ 106-Sg-Seaborgium
- ✓ 107-Bh-Bohrium
- ✓ 108-Hs-Hassium
- ✓ 109-Mt-Meitnerium
- ✓ 110-Ds-Darmstadium
- ✓ 111-Rg-Roentgenium
- ✓ 112-Cn-Copernicium

1 **H** Hydrogen 1.008																	2 **He** Helium 4
3 **Li** Lithium 6.041	4 **Be** Beryllium 9.012											5 **B** Boron 10.81	6 **C** Carbon 12.01	7 **N** Nitrogen 14.01	8 **O** Oxygen 16	9 **F** Fluorine 19	10 **Ne** Neon 20.18
11 **Na** Sodium 22.99	12 **Mg** Magnesium 24.3											13 **Al** Aluminum 26.98	14 **Si** Silicon 28.09	15 **P** Phosphorus 30.97	16 **S** Sulfur 32.07	17 **Cl** Chlorine 35.45	18 **Ar** Argon 39.95
19 **K** Potassium 39.1	20 **Ca** Calcium 40.08	21 **Sc** Scandium 44.96	22 **Ti** Titanium 47.87	23 **V** Vanadium 50.94	24 **Cr** Chromium 52	25 **Mn** Manganese 54.94	26 **Fe** Iron 55.85	27 **Co** Cobalt 58.93	28 **Ni** Nickel 58.69	29 **Cu** Copper 63.55	30 **Zn** Zinc 65.39	31 **Ga** Gallium 69.72	32 **Ge** Germanium 72.61	33 **As** Arsenic 74.92	34 **Se** Selenium 78.96	35 **Br** Bromine 79.9	36 **Kr** Krypton 83.8
37 **Rb** Rubidium 85.47	38 **Sr** Strontium 87.62	39 **Y** Yttrium 88.91	40 **Zr** Zirconium 91.22	41 **Nb** Niobium 92.91	42 **Mo** Molybdenum 95.94	43 **Tc** Technetium 98.91	44 **Ru** Ruthenium 101.07	45 **Rh** Rhodium 102.91	46 **Pd** Palladium 106.42	47 **Ag** Silver 107.87	48 **Cd** Cadmium 112.41	49 **In** Indium 114.82	50 **Sn** Tin 118.71	51 **Sb** Antimony 121.76	52 **Te** Tellurium 127.6	53 **I** Iodine 126.9	54 **Xe** Xenon 131.29
55 **Cs** Cesium 132.9	56 **Ba** Barium 137.3	* 71 **Lu** Lutetium 175	72 **Hf** Hafnium 178.49	73 **Ta** Tantalum 180.95	74 **W** Tungsten 183.85	75 **Re** Rhenium 186.21	76 **Os** Osmium 190.2	77 **Ir** Iridium 192.2	78 **Pt** Platinum 195.08	79 **Au** Gold 196.97	80 **Hg** Mercury 200.59	81 **Tl** Thallium 204.38	82 **Pb** Lead 207.2	83 **Bi** Bismuth 208.98	84 **Po** Polonium 208.98	85 **At** Astatine 209.9	86 **Rn** Radon 222.02
87 **Fr** Francium 223	88 **Ra** Radium 226	** 103 **Lr** Lawrencium 262	104 **Rf** Rutherfordium 261.11	105 **Db** Dubnium 262.11	106 **Sg** Seaborgium 263.12	107 **Bh** Bohrium 264.1	108 **Hs** Hassium 265.1	109 **Mt** Meitnerium 266	110 **Ds** Darmstadtium [271]	111 **Rg** Roentgenium [272]	112 **Cn** Copernicium [277]	113 **Nh** Nihonium	114 **Fl** Flerovium	115 **Mc** Moscovium	116 **Lv** Livermorium	117 **Ts** Tennessine	118 **Og** Oganesson

*Lanthanides

57 **La** Lanthanum 138.91	58 **Ce** Cerium 140.12	59 **Pr** Praseodymium 14.91	60 **Nd** Neodymium 144.24	61 **Pm** Promethium 144.91	62 **Sm** Samarium 150.36	63 **Eu** Europium 151.96	64 **Gd** Gadolinium 157.25	65 **Tb** Terbium 158.93	66 **Dy** Dysprosium 162.5	67 **Ho** Holmium 164.93	68 **Er** Erbium 167.26	69 **Tm** Thulium 168.93	70 **Yb** Ytterbium 173.04

**Actinides

89 **Ac** Actinium 227.03	90 **Th** Thorium 232.04	91 **Pa** Protactinium 231.04	92 **U** Uranium 238.03	93 **Np** Neptunium 237.04	94 **Pu** Plutonium 244.06	95 **Am** Americium 243.06	96 **Cm** Curium 247.07	97 **Bk** Berkelium 247.07	98 **Cf** Californium 251.08	99 **Es** Einsteinium 252.08	100 **Fm** Fermium 257.1	101 **Md** Mendelevium 258.1	102 **No** Nobelium 259.1

Chemistry Memory Work

Unit 4

Gas Laws

1. **Boyle's Law** – At constant temperature, the volume of a gas is inversely proportional to the pressure.
2. **Charles' Law** – At constant pressure, the volume of a gas is directly proportional to the temperature.

Unit 5

Law of Constant Composition – A pure compound always contains the same elements in the same proportions.

Unit 6

Law of Conservation of Mass – In a chemical reaction, the mass of the products is equal to the mass of the reactants.

Avogadro's number: 6.02×10^{23}

Le Chatelier's Principle – If a change is made to a reaction in equilibrium, the reaction will adjust itself to counter the effects of that change.

Unit 8

The Carbon Cycle

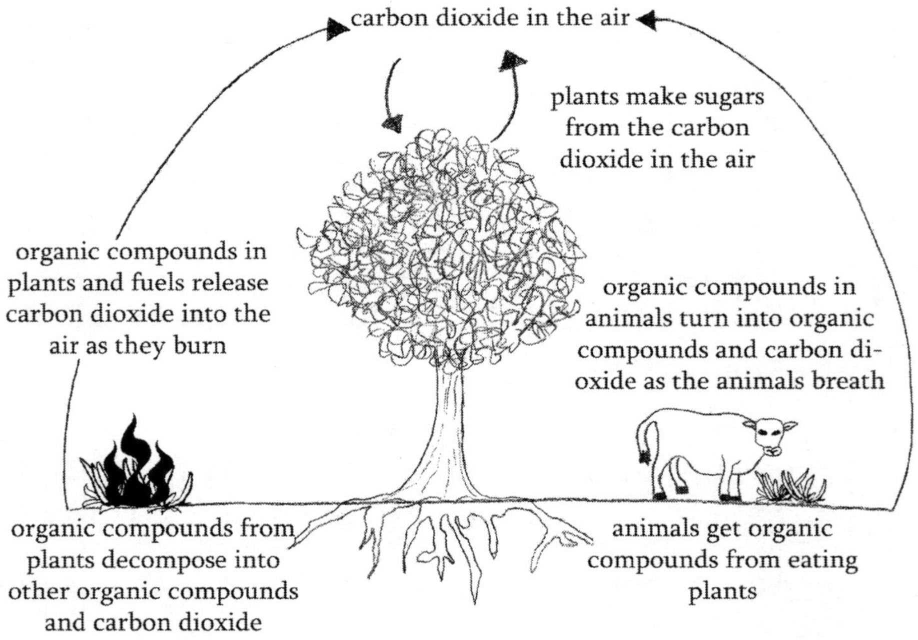

The Four Types of Galaxies

Our universe contains four main classes of galaxy shapes that astronomers know about. The first class is the spiral galaxy. These galaxies have a pinwheel shape with a bulge and thin disk in the center. Spiral galaxies are rich with gas and dust. They contain both young and old stars. Many of these galaxies have been named and cataloged because they are the brightest and easiest to spot in the sky.

The second class of galaxy shapes is the elliptical galaxy. They have a round or oval shape with a bulge in the center, but no disk. Elliptical galaxies have a little cool dust and gas. They contain mostly older stars. Astronomers believe that elliptical galaxies are the most numerous in the universe.

The third type of galaxy shape is the barred spiral galaxy. They have a pinwheel shape with a bar of gas, dust and stars running through the center. Barred spiral galaxies are generally rich in gas and dust. They contain both young and old stars. Some astronomers consider the barred spiral to be a subsection of the spiral class, but most consider the barred spiral to be a separate galaxy shape.

The final type of galaxy shape is the irregular galaxy. These galaxies are shaped just like the name implies, with no regular shape. Irregular galaxies are usually rich with gas and dust. They contain both young and old stars. This class contains a hodgepodge of shapes, basically anything that is not spiral or elliptical in shape.

These are the four fundamental classes of galaxy shapes. Astronomers break each class down into further subclasses to help identify various galaxies. They use the shape of the galaxy and how they believe it was formed to classify each galaxy. Even with all the knowledge we have today, astronomers still have a poor understanding of how a galaxy is formed, so shape remains the best way to identify these starry collections.

*(taken from The Stars and their Stories: A Book for Young People by Alice Mary Matlock Griffith, pg. 9-15, used with permission)

URSA MAJOR AND URSA MINOR

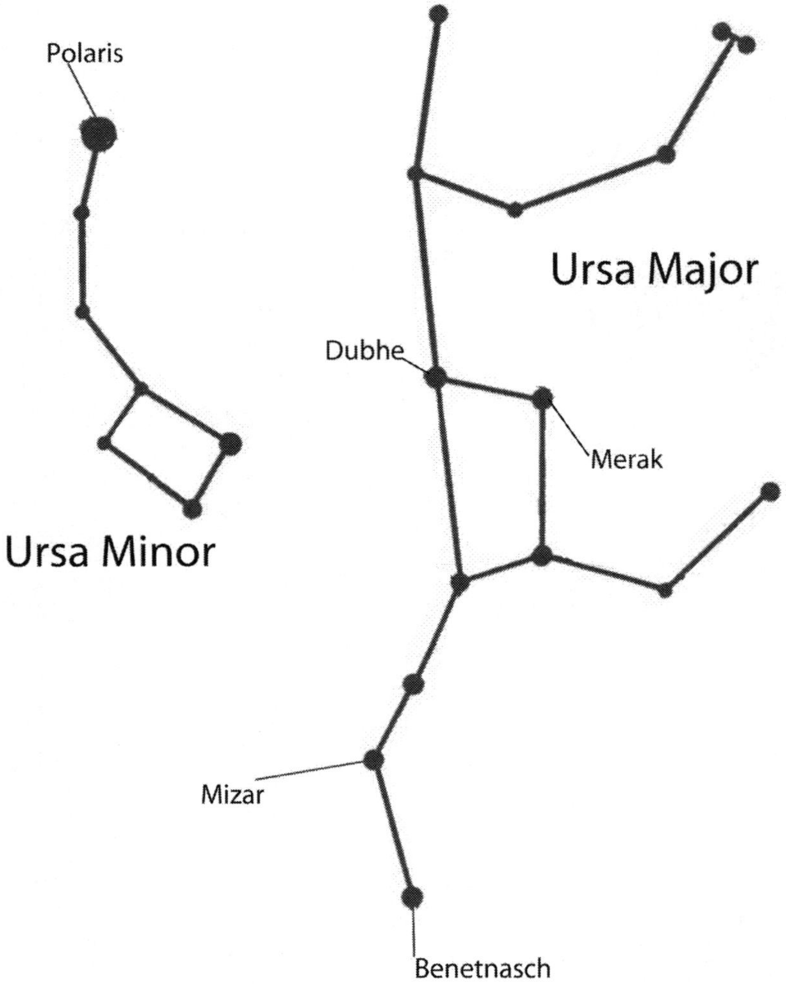

These two constellations whirl so closely to the pole that in our northern latitudes we do not see them set. Each has a group of stars resembling a dipper; and so they are often called the Big Dipper and the Little Dipper.

In the ancient pictures of the Great Bear the legs are much longer than shown here; the two hind legs stretch away to the Lesser Lion, and the left front paw touches the two stars that lie just in front of it here; consequently the figure with its long tail and long legs has small likeness to a bear.

*(taken from The Stars and their Stories: A Book for Young People by Alice Mary Matlock Griffith, pg. 9-15, used with permission)

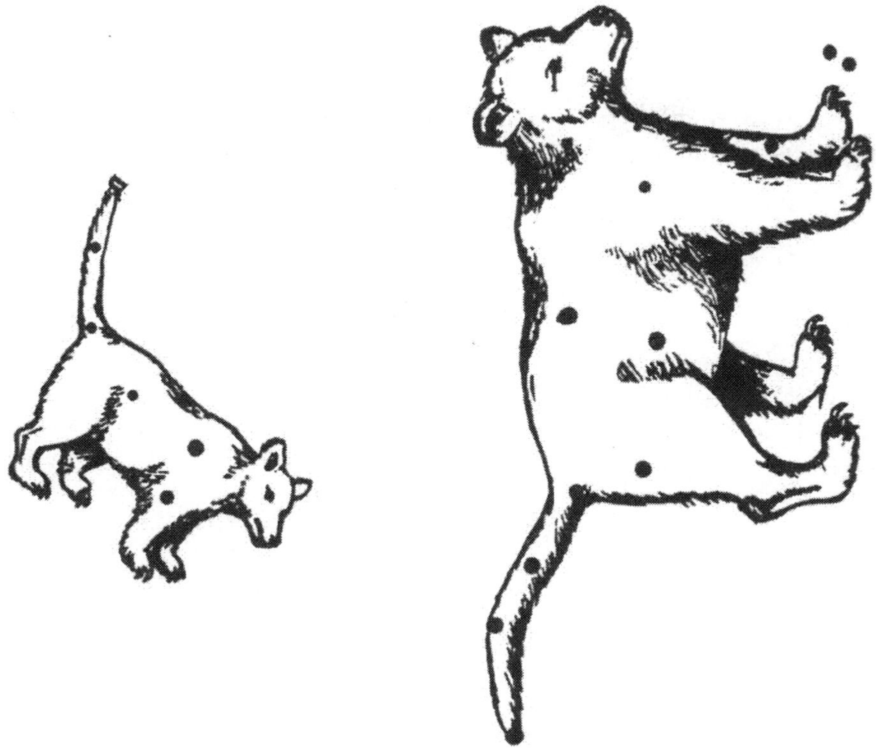

CALLISTO AND ARCAS

The Greeks, if we are to judge from their stories, evidently felt that the friendship of the gods was as likely to bring trouble as their enmity. It proved so in the hard experience of Callisto. She was a beautiful maiden whom Jupiter saw and admired. He showed her many a favor, and visited her often; but he endeavored to keep his visits a secret from his wife, Juno, for she was jealous hearted. The goddess, however, learned of the friendship, and she made up her mind to take a severe revenge upon the human maiden.

Callisto had a baby boy, but not even the love of her son could keep her from going a-hunting. While Callisto was eagerly following the chase one day, she had the bad luck to meet Juno. Now, Juno was the most majestic of all the goddesses, as, to be sure, was entirely befitting in the Queen of Olympus. The mere sight of her must have inspired awe in Callisto. And when the poor lady saw the divine eyebrows straighten into the severity of a frown, and the regal, dark eyes, that Homer loves to describe, glow with the fires of jealousy, she must have felt terror as well as awe. Poor, poor creature, her days of happiness were done. With never a pang of pity, the goddess commanded that she change from a woman to a bear. And lo! the thing was done.

Long, long might the babe wait for his mother, and weep because she returned not.

*(taken from The Stars and their Stories: A Book for Young People by Alice Mary Matlock Griffith, pg. 9-15, used with permission)

How was he to know, or his nurse or his grandfather to know, that far in the lonely forest wandered a she-bear, with the frightened heart of a woman in her breast. All his tears could not change that hairy breast back to the white bosom that had pillowed his baby head so tenderly, could not transform rough paws back to the soft hands that had held him so lovingly. And the bear must have suffered, too. For if Juno was unkind enough to wish for revenge at all, she was doubtless unkind enough not to permit the bear to forget that she had been a woman and a mother. I have sometimes wondered if, when, late in the night, the moon shone quietly and sleepily over hillside and town, the bear-mother did not creep up as close to her old home as she dared, and stand looking sadly at house tops and deserted streets, and long for a sight of her little one; only to steal unsatisfied away in the gray, cold dawn, back to her caves and secret recesses. The story does not say, but I think she must have done so. The pain in her heart, I know, was very great, like that in the heart of the father in Matthew Arnold's "The Forsaken Merman," a poem I like to read to my children, and one that you will like, too, if you will ask your teacher to find it and read it to you.

Young people forget quickly. Callisto's little Areas, if he remembered his mother at all, thought of her as dead. He grew through infancy and well into boyhood, active and strong. Like his mother, he became a great hunter.

Then, when he was fifteen years old, came another unlucky day. He was hunting. Suddenly he perceived near him a bear. Neither had heard the other, and both were surprised. At last, he and his mother had met again. It is possible she saw in him a resemblance to what she had been, or to his father, and recognized him. It is certain that he did not perceive his mother in the shaggy form that stood affrighted before him. For a moment they gazed one upon the other. She had no power to make him understand. And he gave her no time to flee away; but, lifting his bow (bravely, as he would have thought, if he had had time to think about it at all), he was about to send an arrow piercing through her heart.

But the friendship of the gods does not bring only evil. Jupiter had not forgotten the maiden he had once loved so kindly. He could not reunite mother and son by restoring Callisto to human form, for one god may not directly undo what another god has done. He could, nevertheless, do what was still better, since eternal fame is better than a long life, even if life be filled with happiness. What he did do, was, first, to change Areas into a bear, too, and then to transfer both mother and son as stars to the sky. There you will see them as the constellations of the Greater and the Lesser Bears.

One would have thought that Juno would rest satisfied, now that the woman she had feared as a rival had become a group of stars, and could not possibly again give her cause for jealousy. But the vindictiveness in her heart was even yet unsated. If her hated rival was to be placed among the never-fading stars, her revenge must likewise be eternal. She sought out her brother, Neptune, god of the sea, told him her story, and begged him, as a favor to her, to refuse to permit the mother and son ever to enter his realm. Neptune granted her request; and

**(taken from The Stars and their Stories: A Book for Young People by Alice Mary Matlock Griffith, pg. 9-15, used with permission)*

consequently the Great Bear and the Little Bear never sink into the ocean, "the baths of all the western stars."

THE TWO BEARS

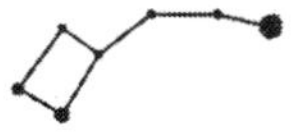

Change as the stars may from night to ight, these two groups, the Great Bear and the Lesser Bear, can be seen from any place in the northern latitudes at any time of the year. We will commence our search by finding them. They are among the easiest to identify, too. Every boy and girl knows the Big Dipper. If you look to the north, you can hardly fail to see it the first thing—seven bright stars set in the shape of a big tin dipper. Four stars make the bowl, and three the handle. Six of these stars are of the second magnitude and one of the third. In the olden times, the Big Dipper was called by some people David's Chariot, the four stars in the bowl representing the wheels, and the three others the horses. Still other people have named them the Seven Oxen, and others again, the Plowshare. In England, Charles's Wain is the popular name. But the name most commonly used is the Great Bear, the stars in the bowl being in the bear's body, and those in the handle being in his tail, as you may see from the chart.

One of the stars in the Great Bear is quite interesting, because it offers an opportunity to see whether you have good, strong eyes or not. The middle star of the three in the tail is named Iizar, and it has a very small companion, called Alcor, "the Test," because it is used as a test of vision, since only a good eye can see it without the help of a telescope.

The two stars that form the outer edge of the bowl of the Big Dipper are called the "Pointers," because they point to the North Star. When we have found the North Star, we have also found the Lesser Bear, because the North Star is the most important star in this group. The Lesser Bear is like the Great Bear in shape, but is smaller. It is sometimes called the Little Dipper. The North Star is the end star of the Lesser Bear's tail, and the end of the dipper handle; and the four stars of the bowl are in his body. The North Star itself is frequently called Polaris, because a line passed through the earth's poles, and extended into the sky, would come very near it.

Within the last few years it has been discovered that Polaris is whirling very rapidly around a dark companion, and that both together are coming towards us at a tremendous speed.

With a knowledge of where the Little and Big Dipper are, and by the aid of the charts, you can easily find all the other constellations in their due times and seasons.

Electrical Circuit Diagram & Instructions

The following electrical circuit will be needed for the week four experiment.

Supplies Needed:

- ✓ Battery (D-Cell or larger)
- ✓ Wire (at least 3 feet)
- ✓ Bare light bulb
- ✓ 2 alligator clips

Instructions:

1. Cut the wire into three lengths.
2. Take one of the wires and attach an alligator clip to one end.
3. Take the next wire and attach one end to the other alligator clip. Then, wrap the other end once around the bottom of a light bulb.
4. Take the final wire and wrap one end around the bottom of the light bulb.
5. Now take the two wires with bare ends and attach those to the two terminals of the battery. (*See the diagram below to visually check your electrical circuit.*)

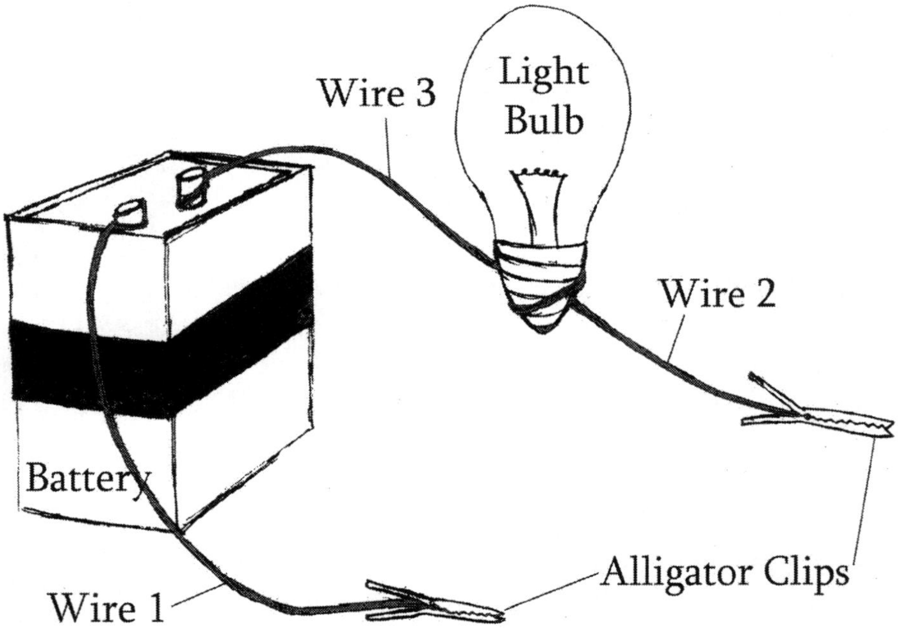

6. The electrical circuit is now ready for testing. Simply clip the alligator clips on either side of your sample to complete the circuit. If the light bulb lights up, the material conducts electricity; if it does not, the sample does not conduct electricity.

Activity Log

Activity	Date
What I did/saw/learned	

Activity	Date
What I did/saw/learned	

Activity	Date
What I did/saw/learned	

Activity Log

Activity	Date
What I did/saw/learned	

Activity	Date
What I did/saw/learned	

Activity	Date
What I did/saw/learned	

Activity Log

Activity	Date
What I did/saw/learned	

Activity	Date
What I did/saw/learned	

Activity	Date
What I did/saw/learned	

Activity Log

Activity	Date
What I did/saw/learned	

Activity	Date
What I did/saw/learned	

Activity	Date
What I did/saw/learned	

Glossary

Glossary

A

- **Acid** – A hydrogen containing compound that splits in water to give hydrogen ions.
- **Acidic Solution** – A solution that contains an acid, or a solution that has a pH less than seven.
- **Adhesive** – A sticky substance used to join two surfaces together.
- **Alcohol** – A series of organic compounds in which a hydroxyl group is bound to a carbon atom; they all have the general formula $C_nH_{n+1}OH$.
- **Alkali** – A base that dissolves in water to form a hydroxide ion.
- **Alkaline Solution** – A solution that contains a base, or a solution that has a pH greater than seven.
- **Alloy** – A mixture of two or more metals, or a metal and a nonmetal.
- **Anabolism** – The synthesis of complex molecules from smaller ones; occurs in living things and is known as creative metabolism.
- **Anode** – The positively charged electrode by which current enters the cell.
- **Asteroid** – Large chunks of rock and metal that orbit the sun in the region between Mars and Jupiter.
- **Atom** – The smallest particles of an element that retains the chemical properties of the element.
- **Atomic Mass** – The average mass number of the atoms in a sample of an element.
- **Atomic Number** – The number of protons in the nucleus of an atom.
- **Axis** – An imaginary line through the center of a planet, around which it rotates.

B

- **Base** – A compound that reacts with an acid to produce water and a salt.
- **Black hole** – Formed by a collapsed star, has a very strong gravitational pull.
- **Boiling Point** – The temperature at which a substance turns from a liquid into a gas.
- **Buffer** – A solution that resists changes in pH.

C

- **Carbohydrates** – A group of organic compounds that consists of carbon, hydrogen and oxygen.

- **Catabolism** – The breaking down of complex molecules into smaller ones; occurs in living things and is known as destructive metabolism.
- **Catalyst** – A substance which speeds up a chemical reaction without being changed by the reaction.
- **Cathode** – The negatively charged electrode by which current leaves the cell.
- **Chemical Compound** – A substance made up of two or more different elements that are chemically joined in fixed proportions.
- **Chemical Reaction** – An interaction between the atoms of two substances where the atoms rearrange to form two new molecules.
- **Chromatography** – A method of separating the substances in a mixture by the rate they move through or along a medium, such as filter paper.
- **Cleavage** – The splitting of a crystal along a certain plane.
- **Cluster** – A group of galaxies that are found close together.
- **Comet** – A chunk of frozen gas and dirt that has an orbit.
- **Compound** – A substance that consists of a combination of different elements.
- **Concentration** – A measure of the strength of a solution, i.e. the amount of the solute in the solvent.
- **Conductivity** – The measure of a substance's ability to conduct heat or electricity.
- **Constellation** – The pattern that a group of stars seems to make in the sky.
- **Covalent Bonding** – A chemical bond between two atoms, in which each atom shares an electron.
- **Crater** – The hole formed in a planet's surface by the impact of a meteorite.

D

- **Density** – A measure of the amount of matter in a substance compared to its volume.
- **Detergent** – A synthetic or organic substance, not prepared from a fat, that helps to remove dirt, grease and oil in water.
- **Diffusion** – The spreading out of a gas to fill the available space of the container it is in.
- **Dissociation** – The process by which a substance is split up into its ions in a solution.
- **Dwarf planet** – A minor planet in the solar system.
- **Dye** – A water-soluble substance used to color materials.

E

- **Eclipse** – When one celestial body casts a shadow on another celestial body.
- **Elasticity** – The ability of a substance to stretch and then return to its original shape.

- **Electrolyte** – A substance that conducts electricity when it is in solution.
- **Electron Shell** – A region around the nucleus of an atom where a specific number of electrons can exist.
- **Element** – A substance made up of one type of atom, which cannot be broken down by chemical reaction to form a simpler substance.
- **Emulsion** – Tiny particles of one liquid dispersed in another liquid, such as oil and water.
- **Endothermic Reaction** – A chemical reaction that takes in heat.
- **Enzyme** – A catalyst used by living things to increase the speed of a natural chemical process.
- **Equilibrium** – The point at which the rate of the forward reaction is equal to the rate of the backward reaction.
- **Exothermic Reaction** – A chemical reaction that gives off heat.

F

- **Fermentation** – A chemical reaction in which a sugar is broken down by yeast or an enzyme to yield alcohol and carbon dioxide.
- **Freezing Point** – The temperature at which a substance turns from a liquid into a solid.

G

- **Galaxy** – A body held together by gravity that is made of millions of stars, gas, and dust.
- **Galilean moons** – Jupiter's 4 largest moons, Io, Ganymede, Callisto and Europa, they are named after Galileo, who discovered them.
- **Gas giant** – A large planet that is mostly composed of gas.
- **Group** – A column of elements in the periodic table that have similar properties, electron configurations and valencies.

H

- **Hydrocarbon** – A chemical compound that consists of only hydrogen and carbon.

I

- **Immiscible** – Liquids that cannot be blended together.
- **Indicator** – A substance that changes color as the pH of the solution changes.
- **Ion** – An atom or group of atoms that has become charged by gaining or losing one or more electrons.
- **Ionic Bonding** – A strong chemical bond that is formed by the attraction between two ions

of opposite charges.

- **Irreversible Reaction** – A reaction in which the products can never be turned back into the reactants.
- **Isomer** – A compound that contains the same number of atoms as another compound, but in a different arrangement.

J

K

- **Kinetic Theory** – As the temperature rises, particles move faster and therefore take up more space.

L

- **Lattices** – The regular arrangement of repeating patterns of atoms, molecules or ions in a solid structure.
- **Lipids** – A group of solid esters that are stored in living things as a source of reserve energy, also known as fats.

M

- **Malleable** – A characteristic of a metal that means it is bendable and easily shaped.
- **Metal** – The largest class of elements, usually they are shiny and solid at room temperature.
- **Metallic Bonding** – A chemical bond where positive metal ions form a lattice structure with freely moving electrons between them.
- **Meteor** – A meteoroid that starts to burn as it enters the atmosphere.
- **Meteoroid** – A small piece of space debris.
- **Minerals** – A group of inorganic compounds that are an essential components to the body's chemical processes.
- **Miscible** – Liquids that can be blended together.
- **Mixture** – A combination of two or more elements or compounds that are not chemically combined.
- **Molecule** – A substance that is formed when two or more atoms chemically join together.
- **Moles** – The SI unit for the amount of a substance.
- **Monomers** – Small molecules that can join together to form long-chain polymers.
- **Moon** – A celestial body in orbit around a planet.

N

- **Natural Fiber** – A fiber made from plants or animals, such as wool.
- **Natural Polymer** – A polymer that is produced by a living organism, such as cellulose, starch, chitin and proteins.
- **Nebulae** – Clouds of dust and gas found in space.
- **Neutralization** – The process by which you make a solution neither acidic nor alkaline.
- **Nonmetal** – A class of elements that typically forms negative ions; they are usually dull solids or gases.

O

- **Orbit** – The path of one celestial body around another.
- **Organic Compound** – A compound that contains the element carbon.
- **Oxidation** – Occurs when an atom loses electrons in a reaction, or when a substance loses a hydrogen or gains an oxygen atom in a chemical reaction.

P

- **Period** – A set of elements that have the same number of electron shells, shown as rows in the periodic table.
- **pH** – A measure of the acidity or alkalinity of a solution.
- **Photosphere** – The surface of the Sun.
- **Pigment** – A substance that gives color to the surface of a material.
- **Planet** – A large globe composed of rock, liquid, or gas that revolves around a star.
- **Planetarium** – A device used to project images of the stars or depict the solar system.
- **Plasma** – The fourth state of matter that exists only at very high temperatures, such as those found on the sun.
- **Polymer** – A substance with a long-chain of molecules, formed from smaller molecules called monomers.
- **Poor Metal** – A group of metals that are soft and weak.
- **Pressure** – The amount of force pushing on a given area.
- **Product** – A new substance produced in a chemical reaction.
- **Prominence** – Large eruptions or loops of flaming gas from the Sun's surface.
- **Protein** – A natural polymer that is made up of amino acids.

Q

R

- **Raw Materials** – The basic materials used to make a product.
- **Reactant** – The substance present at the beginning of a chemical reaction.
- **Reactivity** – The tendency of a substance to react with other substances.
- **Redox Reaction** – A reaction that primarily involves the transfer of electrons between two substances.
- **Reduction** – Occurs when an atom gains electrons in a reaction, or when a substance gains a hydrogen or loses an oxygen atom in a chemical reaction.
- **Reversible Reaction** – A reaction in which, under the right conditions, the products can be turned back into the reactants.

S

- **Salt** – A type of compound that is formed when an acid and a base react.
- **Semiconductor** – A substance that only conducts electricity under certain conditions.
- **Semimetal** – A group of elements that have characteristics of both metals and nonmetals.
- **Slag** – Impurities found in iron ore that are leftover from the smelting process.
- **Soap** – A metallic salt derived from a fat, that is used to help remove dirt, grease and oil.
- **Solar wind** – A constant stream of particles that flow into space from the Sun.
- **Solubility** – The ability of a solute to be dissolved.
- **Solute** – The substance that dissolves in the solvent to form a solution.
- **Solution** – A homogenous mixture of two or more substances.
- **Solvent** – The substance in which the solute dissolves to form a solution, typically a liquid.
- **Star** – A massive, hot, shining ball of gas.
- **Starch** – A natural polymer that is made up of glucose monomers.
- **Sublimation** – Occurs when a substance changes directly from a solid to a gas without changing into a liquid.
- **Sunspot** – Dark, cooler areas of the Sun's photosphere.
- **Supercluster** – The largest structure in the universe, composed of many galaxy clusters.
- **Synthetic Fiber** – A fiber made from petrochemicals, such as nylon.
- **Synthetic Polymer** – A polymer that is man-made, also known as plastics.

T

- **Temperature** – A measure of the heat that a substance contains.

- **Titration** – A method of finding the concentration of acidic or alkaline solutions.

U

- **Universe** – The collection of all the matter, space, and energy that exists, also known as the cosmos.

V

- **Vitamins** – A group of organic compounds that are essential for the normal growth and nutrition of living things.
- **Volume** – The space occupied by matter.

W

X

Y

- **Yeast** – A single-celled fungus used in fermentation.

Z

- **Zodiac** – The twelve constellations through which the sun, moon, and planets appear to move.

Made in United States
Orlando, FL
04 November 2026

90209477R00190